I0469729

Disclaimer

BLAST-RESISTANCE BENEFITS OF SEISMIC DESIGN

Phase 2 Study: Performance Analysis of Structural Steel Strengthening Systems

Robert G. Pekelnicky

Stanley C. Woodson

Steven C. Sweeny

John R. Hayes, Jr.

Joe Magallanes

David R. Bonneville

Chris D. Poland

FEMA P-439B

November 2010

Preface

The Federal Emergency Management Agency (FEMA), which is part of the Department of Homeland Security (DHS), and the National Earthquake Hazards Reduction Program (NEHRP) both have a similar goal, which is to encourage design and building practices that address the earthquake hazard and minimize the resulting risk of damage and injury. A related FEMA goal is to present guidance that addresses all hazards in a coordinated manner. This publication is the second and final in a series developed with this related goal in mind, and examines the relationship between seismic resistant design and blast resistant design and attempts to quantify the blast resistance benefit a building designed to withstand high seismic loads would inherently incorporate.

This series of publications was developed in response to the September 11, 2001 terrorist attacks on the New York World Trade Center and the subsequent events that led to the formation of DHS and an increased emphasis on preparedness and mitigation of terrorism-related hazards. One issue that FEMA began shortly after that was to examine whether lessons learned in response to natural hazards could be effectively applied to protect building occupants from human threats. Important similarities between seismic and blast loadings (e.g., both can impose extreme horizontal forces on a structure within a small time frame) lend themselves to such examination.

The first publication of this series is *Blast Resistance Benefits of Seismic Design Phase 1 Study: Performance Analysis of Reinforced Concrete Strengthening Systems Applied to the Murrah Federal Building Design* (FEMA 439A, December 2005). This publication was developed based on data from the bombing of the Alfred P. Murrah Federal Building in Oklahoma City in April 1995. That event was documented in *The Oklahoma City Bombing: Improving Building Performance Through Multi-Hazard Mitigation* (FEMA 277, August 1996). One finding of that report was that "…application of mitigation strategies developed for FEMA for wind and earthquake can significantly improve blast resistance." That conclusion was based on the fact that, although not required by Oklahoma City building codes either then or now, had the Murrah Building been designed as a concrete moment frame with seismic resistance features such as structural detailing and structural system redundancy, the progressive collapse could have been avoided. The Phase 1 FEMA 439A report demonstrated that, with such seismic design features in place, the structural system would have been better able to dissipate and manage the blast load effects, reducing or avoiding catastrophic chain-reaction impacts on portions of the building that were not destroyed as a direct result of the bomb blast.

This publication, *Blast Resistance Benefits of Seismic Design Phase 2 Study: Performance Analysis of Steel Frame Strengthening Systems* (FEMA 439B, November 2010), supplements the first phase study by using the same study scenario, but with a steel frame building. A federally owned steel frame building located in a low seismic area was selected from the GSA inventory and a series of seismic strengthening designs were developed based on the original plans. The original building and the seismically strengthened designs were then evaluated using the same blast characteristics and modeling used in the Phase 1 Murrah Federal Building study. The results were even more encouraging than the first study, and showed that, at least for this one example, a steel frame building demonstrated a significant amount of resistance and redundancy.

Both of these studies were performed under an agreement with the U.S. Army Engineer Research and Development Center (ERDC), Construction Engineering Research Laboratory (CERL) to conduct advanced structural analyses on specific examples of different construction types in order to quantify the blast resistance benefits afforded by appropriately designed seismic resistance features.

The purpose of these studies is not to develop or support a hypothesis that seismic design is equivalent to blast-resistant design. It is a given that adequate blast resistance can be achieved only through building specific analysis and design for that express purpose. However, the findings of these studies strongly indicate that a building originally designed or later upgraded to address high seismicity will also provide a significant level of blast resistance. Although cost constraints can be daunting for building owners who need to achieve higher levels of protection than conventional construction methods provide, the findings of this study should encourage an owner to consider the potential cost benefits of addressing both types of structural safety hazard in a coordinated or holistic way. Considering the significant blast resistance benefits offered by seismic design, the building owner may find that achieving complete blast protection requires only an incremental cost increase over some types of seismic strengthening described in this report.

The analyses and narrative for this report were developed by Steven C. Sweeney, ERDC-CERL; Stanley C. Woodson, ERDC Geotechnical and Structures Laboratory (ERDC-GSL); David R. Bonneville, Chris D. Poland and Robert G. Pekelnicky, Degenkolb Engineers; Joe Magallanes, Karagozian and Case; and John R. Hayes, Jr., formerly ERDC-CERL and now with the NEHRP at National Institute of Science and Technology (NIST). A technical oversight panel was formed for the project, to bring in expertise on the behavior of structural steel in extreme loading environments. Panel members were Dr. John Barsom, Barsom Consulting; Mr. Charles Carter, American Institute of Steel Construction; Mr. John Crawford, Karagozian and Case; Dr. Ted Galambos, the University of Minnesota; Mr. Harold Sprague, Black and Veatch; and Dr. Andrew Whittaker, the University at Buffalo. FEMA gratefully acknowledges the assistance provided by Rosa Bertino and Tammy Chang of Degenkolb Engineers in finalizing this publication and for the American Institute for Steel Construction (AISC) and Kargozian and Case, who funded and performed a blast test of a representative steel column member to verify its performance for this study. The FEMA Project Officer was Michael Mahoney and the FEMA Technical Advisor was Robert Hanson. FEMA also wishes to thank Bela Palfalvi, General Services Administration (GSA) San Francisco, who provided valuable assistance in gathering as-built drawings of the subject building.

Table of Contents

List of Appendices

List of Figures

List of Tables

Executive Summary

In the aftermath of the September 11, 2001 terrorist attacks on the New York World Trade Center and the Pentagon, the Federal Emergency Management Agency (FEMA), which had been incorporated into the newly formed Department of Homeland Security (DHS), focused an increased emphasis on preparedness and mitigation of terrorism-related hazards. One particular study that FEMA undertook shortly after that time was to examine whether lessons learned in response to natural hazards could be effectively applied to protect building occupants from human threats. In particular, important similarities between seismic and blast loadings (e.g., both can impose extreme horizontal forces on a structure within a small time frame) lent themselves to such examination.

An earlier event, the 1995 bombing of the Alfred P. Murrah Federal in Oklahoma City, had already emphasized the urgency of developing ways to prevent catastrophic structural failures when buildings are subjected to a bomb blast. The FEMA report *The Oklahoma City Bombing: Improving Building Performance Through Multi-Hazard Mitigation* (FEMA 277, August 1996) on the impact to the Murrah Building as a result of the Oklahoma City blast hypothesized that details used in structures to resist the effects of damaging earthquakes could have reduced the damage to the Murrah Building. Earthquake resistant design and detailing provisions are commonplace in structural engineering today. If these seismic resistant design methods could also be shown to provide some level of improved blast and collapse resistance for structures, the overall cost of providing an adequate level of blast and collapse resistance could be significantly reduced, making this type of protection more affordable and attractive for building owners.

Beginning in 2002, FEMA funded a study to examine the hypothesis that seismic design provisions could also help to provide some quantifiable level of blast resistance. The first phase of that study focused on the Murrah Building under the 1995 bombing scenario, with seismic upgrade and redetailing schemes used in areas of high seismicity. The building was evaluated as if it was located in a zone of high seismicity and three seismic upgrade schemes were designed for the high seismicity of San Francisco. Additionally, the exterior frame was redetailed as a seismically special moment resisting frame. Those four modified building designs were then subjected to an analysis using the 1995 Murrah Building bombing scenario. The blast analyses showed that the two upgraded structures with the new seismic force resisting elements placed on the building's exterior and the redetailed structure had considerably greater resistance to the blast effects and post-blast collapse. The upgraded structure that had new seismic shear wall elements placed within its interior saw no appreciable change in its blast response and collapse resistance. That study is reported in FEMA's *Blast Resistance Benefits of Seismic Design Phase 1 Study: Performance Analysis of Reinforced Concrete Strengthening Systems Applied to the Murrah Federal Building Design* (FEMA 439A, December 2005).

Since the FEMA 439A study focused specifically on the Murrah Building, a 1970s reinforced concrete ordinary moment frame gravity-load-resisting system supplemented with reinforced concrete shear walls to resist lateral forces, interpolation of the studies conclusions to other building types would not be prudent without additional studies. Therefore, FEMA commissioned a subsequent study to examine the effects of seismic design on a structural steel building. That study is the subject of this report.

This study looked at a structural steel building designed and constructed during the same era that was similar to the Murrah Building, a mid-rise federal office building situated in an area with low seismic hazard. A six-story steel framed building constructed in 1970 was chosen for the study. That building's lateral force resisting system is composed of welded steel moment frames along each column line. The building, unlike the Murrah Building, is very regular in plan.

Like the Murrah Building in the previous study, the steel building was artificially re-sited to San Francisco, an area of very high seismicity. The building was evaluated based on all three tiers of evaluation in ASCE 31-03. All three tiers indicated that the building was seismically deficient for the San Francisco site. The main issues related to the moment frame connection and column splice details not being adequate for the demands placed on them as the beams yield and as the building has excessive frame drift in response to the large seismic demands.

Three strengthening schemes were developed to mitigate the identified seismic deficiencies. The first scheme involved adding bucking-restrained braced frames within select interior bays to strengthen and stiffen the structure. The second scheme upgraded all moment frame connections and column splices to provide robustness for the beams to deform inelastically. The final scheme involved adding buckling-restrained frames along the perimeter of the structure. In conjunction with that scheme, a hat-truss was added around the top story perimeter so that every exterior column could hang if its first story was damaged. This was done to provide a "smart scheme" which sought to address both seismic hazards and blast hazards.

As with the Murrah Building, the perimeter frame of the steel building was also redetailed to conform to current Special Moment Frame detailing requirements. This involved re-orienting the columns and increasing some of their sizes to satisfy the strong column-weak beam seismic criterion. Different column splice connections were also employed.

The original building, the three upgraded schemes, and the redetailed structure were all then subjected to the Murrah Building blast scenario. Unlike the Murrah Building, however, there was no baseline performance for comparison. Therefore, the following study was conducted to estimate this initial blast performance. Initially, single degree of freedom models were used to estimate the blast response. Because of the uncertainty and lack of validation of those models for close-in blasts, some high-fidelity finite element analyses were performed on a portion of the structure. An actual physical blast test of a similar column was also carried out in conjunction with this study. These analyses, coupled with the test and engineering judgment of the project team, led to a baseline response estimate that had the splice of the column closest to the blast failing, the second and third floor girders in the two bays closest to the bomb significantly damaged, and the second floor slab one bay each side of the bomb completely destroyed.

The upgraded structures showed varying improvement to the direct blast response. The interior scheme and the re-detailed frame performed similar to the baseline case. The connection upgrade scheme showed slight improvement, while the exterior frame scheme showed the greatest improvement.

The post-blast progressive collapse assessment was subject to more uncertainties than the direct blast evaluation. Because of that, best and worst case scenarios were postulated based on analysis and engineering judgment. The best-case solution was that no collapse occurred following the loss of members due to direct blast for all structures. The worst case scenario had the two bays adjacent to the blast collapsing the entire height of the structure for the original building and the interior scheme, while the connection upgrade, "smart scheme" and re-detailed scheme showed no collapse.

The study results indicate that if the best-case scenario holds true, then the existing steel structure was highly resistant to post blast collapse for every configuration from the original building to the smart scheme. If the worst-case scenario were to occur, then the study shows that seismic upgrade schemes and seismic detailing, when applied to the perimeter structural elements, can provide increased resistance to blast and progressive collapse.

Additionally, several observations were made about the positive impact that regular plan configurations with substantial redundancy can have on a structure's blast and collapse resistance. Since the Murrah Building was configured to be two bays wide, the collapse of the front caused the loss of approximately half of the floor area. Since the building in this study was configured to be six bays deep, the loss of the entire front frame would only result in a 17% loss in floor area. Also, this building, unlike the Murrah Building, had a complete space frame without any discontinuous columns and transfer girders.

The study also highlighted numerous uncertainties and research needs for steel structures subjected to direct blast effects and their post-blast collapse resistance. There is limited blast testing of steel elements, particularly sub-assemblages where connection behavior under blast conditions can be benchmarked. Further research is needed to have physical tests to correlate with high fidelity finite element analysis and the single-degree-of-freedom models commonly used in blast response evaluation.

1   Introduction

1.1   *Background*

In 1995, terrorists attacked the Alfred P. Murrah Federal Building in Oklahoma City, OK, using a truck bomb composed of an ammonium nitrate fuel oil (ANFO) mixture that was estimated to contain an explosive yield equivalent to approximately 4,000 lb of TNT (Reference 1). The bomb was detonated approximately 15.6 feet from a critical column on the street face of the building, destroying the column and several other adjacent structural elements and triggering a progressive collapse mechanism that involved almost half of the floor area of the building. The term *progressive collapse* is used to describe a chain reaction of structural failures that occurs following damage to a relatively small portion of a structure; the damage that ultimately results is disproportionately large if compared with the direct change that starts the chain reaction.

The Murrah Building was an older reinforced concrete moment frame structure that was designed and constructed in the 1970s. Because Oklahoma City is located in a region of relatively low seismic activity, the original structural design did not consider the effects of earthquake-induced ground motions in accordance with the applicable building code provisions for that time and location.

An analysis of the damage to the building was reported by the Building Performance Assessment Team (BPAT) in Federal Emergency Management Agency (FEMA) Report 277, *The Oklahoma City Bombing: Improving Building Performance Through Multi-Hazard Mitigation* (Reference 1). In the report, BPAT members hypothesized that many of the techniques used to increase the earthquake resistance of buildings can also improve blast resistance and progressive collapse resistance.

Since the 1995 Murrah Building attack, engineers in the United States and abroad have discussed the possible parallels between seismic design and blast- and progressive collapse-resistant design. The question that arises is whether a building that has good earthquake resistance will resist blast and progressive collapse damage more effectively than one that does not.

To begin addressing the question using analytical data, FEMA commissioned a two-phase study. The overall purpose of the study was to assess:

- What relative improvements in blast- and progressive-collapse resistance accrue in older buildings when new strengthening measures to enhance earthquake response are undertaken; and,

- What relative improvements in blast- and progressive-collapse resistance are present in new buildings that are constructed with current building code-required seismic *detailing* as compared with buildings that have not been designed to provide significant seismic resistance? *Detailing* in this case was the process of designing into structural elements the features needed to ensure that they perform well during any designed loading event.

Phase 1 of this study used the original Murrah Building to focus on assessing reinforced concrete buildings. FEMA Report 439A, *Blast-Resistance Benefits of Seismic Design, Phase 1 Study: Performance Analysis of Reinforced Concrete Strengthening Systems Applied to the Murrah Federal Building Design* (Reference 2), thoroughly documents that assessment. The report concludes that improvements in blast- and progressive collapse-resistance can accrue both from seismic strengthening of older reinforced concrete buildings and from using seismic detailing that is required for reinforced concrete buildings that are constructed in areas of high seismic activity. However, the report also notes that direct conclusions about construction materials and systems other than reinforced concrete should not be drawn without further study.

FEMA then commissioned Phase 2 of this study, which is described in this report. In Phase 2, an older structural steel moment frame was examined in essentially the same manner as that documented in FEMA 439A. A steel moment frame that was constructed at roughly the same time as the Murrah Building was examined for its response to a bombing of the same size and location with respect to critical structural elements. The building's response in its original configuration and in several seismically strengthened configurations was determined through analysis. Using the analyses that were performed, engineering observations regarding the performance of structural steel buildings were then made.

1.2   Objectives

This study is part of an investigation as to whether the effectiveness of strengthening measures intended to improve a building's resistance to earthquake motions can also improve the resistance of a building to blast effects and progressive collapse. This phase of the study addresses the seismic strengthening of older steel moment frame buildings. In an approach that is very similar to the evaluation documented in FEMA 439A, a specific existing steel moment frame building was used to exemplify steel moment frame behavior, and the results of that one building are then generalized to the class.

A key aspect of the study was establishing the "baseline" performance of the building that is evaluated. This "baseline" was the postulated response of the unimproved (as-built) building in the established Murrah Building truck bombing scenario (Figure 1-1). In contrast to the known response of the Murrah Building that was used to validate the structural modeling procedures that are documented in FEMA 439A, the blast and progressive collapse response of the original steel structure had to be established through analysis.

In addition to the baseline performance analysis, the study focused on the building's response if it were to be upgraded to improve its earthquake resistance. The first step of this aspect of the project was a structural seismic evaluation of the building, assuming it is located as built in an area of high seismic activity, a seismic design category (SDC) E as defined by IBC 2003 (Reference 3) and IBC 2006 (Reference 38). The evaluation was followed by designing strengthening measures for the building to mitigate seismic deficiencies identified in the evaluation. The strengthening designs provided sufficient structural information to support analysis of the blast and progressive collapse response of the upgraded building at the same depth that was reported in FEMA 439A, and to develop construction cost estimates. The blast response of the strengthened building used the same blast scenario reported in FEMA 277 (Reference 1) and FEMA 439A. Based on those analyses, conclusions were drawn about the potential contribution of seismic strengthening to blast and progressive collapse resistance in the subject building, and, potentially, steel moment frame buildings in general.

In addition to designing strengthening schemes for the building, the project included the step of seismic re-detailing the steel moment frame of the original building as a special moment frame that fully complies with current seismic building codes, without specifically designing frame elements for lateral forces that would result from earthquake motions in high seismic areas. This re-detailed frame is analyzed for its blast response in the same manner used for the seismic strengthening schemes. This re-detailing investigation is similar to that undertaken for the concrete system in the FEMA 439A report.

1.3 Approach

The first step in the project involved the selection of a satisfactory building for the study. At the request of the project team, the General Services Administration (GSA) surveyed its existing building inventory for steel moment frame buildings that were constructed during the early to mid 1970s in areas of relatively low seismic activity. GSA provided a list of several candidate buildings. The building that was selected for this study was constructed in the eastern U.S. in 1970. Because the building is still in the active GSA building inventory, it is not specifically identified in this study. Specific identifying features are not shown, and alterations have been made in the building floor plan. In general, however, the "as-built" conditions of the actual building, including specific structural details, have been used. Section 1.4 provides a summary description of the structural system. Figures 1-2 through 1-12 show key features of the building as it was studied.

The study addresses the effects of seismic strengthening measures on blast resistance. To gain the clearest indication of their effectiveness, strengthening measures were designed for high seismicity conditions for a building that was originally constructed for low seismicity conditions. This approach establishes whether there is a substantive link between seismic strengthening and improved blast resistance than would be obtained using a lower seismic demand. Possible future studies could examine the effects that the less substantial structural improvements associated with more moderate seismic demand might provide. This approach is consistent with that used to develop FEMA 439A. Site characteristics are discussed in Chapter 2.

Degenkolb Engineers performed seismic screening and evaluations using procedures outlined in American Society of Civil Engineers (ASCE) 31-03, *Seismic Evaluation of Existing Buildings* (Reference 4), which is an update of FEMA 310, *Handbook for the Seismic Evaluation of Buildings – A Prestandard* (Reference 5). A Tier 1 screening was performed, followed by Tier 2 and Tier 3 evaluations. The Tier 2 evaluation incorporated a linear dynamic procedure (LDP) structural analysis and the Tier 3 evaluation used a nonlinear static procedure (NSP) structural analysis. The evaluations found a number of critical seismic deficiencies. The screening and evaluations are summarized in Chapter 2 and explained more fully in Appendix A.

Following the seismic screening and evaluations, three strengthening schemes were developed for the building. The structural designs and analyses were performed in general conformance with the requirements of FEMA 356, *Prestandard and Commentary for the Seismic Rehabilitation of Buildings* (Reference 6).

The first two strengthening schemes were developed with no consideration for blast or progressive collapse resistance. Scheme 1 involves adding buckling-restrained braced frames (BRBFs) along interior column lines in both major axis dimensions of the building, increasing column capacities to accommodate the increased forces induced by the BRBFs, increasing selected first floor beam capacities, and improving the foundation system by adding piles. Scheme 2 involves upgrading all of the existing moment-resisting connections using the Welded Bottom Haunch (WBH) connection that is described in Section 6.6.2 of FEMA 351, *Recommended Seismic Evaluation and Upgrade Criteria for Existing Welded Steel Moment Frame Buildings* (Reference 7), and improving column splices.

Scheme 3 was termed a "smart scheme," in that it considers the possibility of column losses at lower levels in the building, along with seismic strengthening. The "smart" aspect of the scheme entails adding a "hat truss" system at the roof level on the four exterior frame lines, to supplement the existing gravity load system and provide an alternate load path if an existing column should be lost (such as due to terrorist attack). BRBFs are again used, but they are added on the exterior frame lines of the building. As in Scheme 1, existing column and beam capacities are increased where needed.

Details of strengthening Schemes 1-3 are summarized in Chapter 2 and discussed fully in Appendix C.

In addition to the three strengthening schemes, a fourth approach, designated as Scheme 4 for consistency, involves upgrading all perimeter structural member details to meet the minimum new construction detailing requirements for special moment frame members in ANSI/AISC 341-02, *Seismic Provisions for Structural Steel Buildings* (Reference 8). This effort does not involve a redesign of the original members for higher specified force and drift levels. Column sizes are increased to provide "weak beam – strong column" behavior, and selected columns are reoriented so their strong bending axes are in the planes of the respective exterior frames. Beam connections to the columns are re-designed as reduced beam sections, or "dog-bone" connections, as described in Section 3.5.5 of FEMA Report 350, *Recommended Seismic Design Criteria for New Steel Moment-Frame Buildings* (Reference 9). Scheme 4 special moment frame details are summarized in Chapter 2 and discussed fully in Appendix C.

In addition to the original structural system, the four different new structural system designs (three strengthening schemes and one re-detailed version of the original frame system) were subjected to blast response analyses. The initial intent was to use only SDOF models in the blast response analyses, similar to the approach reported for the Murrah Building in FEMA 439A. The SDOF models used software that was developed for designing blast-resistant facilities. The software has been validated using the results of a select number of tests and field observations dating back to World War II. Each blast response analysis used essentially the same bomb scenario that was reported in FEMA 277.

Complementing the SDOF blast response modeling, several nonlinear finite element analyses (NLFEA) were conducted. The NLFEA incorporated numerical models that have been developed in support of ongoing field blast testing of structural steel components in a project that is jointly funded by the Defense Threat Reduction Agency (DTRA) and GSA. The models used in the NLFEA have been validated through comparison with field test results, so they provided useful benchmarks for gaging the validity of the SDOF modeling results.

Chapter 3 presents the results of the blast response analyses for each of the four systems (three upgrade schemes, one re-detailed scheme). Both the SDOF modeling results and the NLFEA results are shown. The blast response analyses are discussed more fully in Appendix D.

The blast response analysis results were then used to project the degree of progressive collapse that could occur for each of the four strengthening schemes. These projections were based on the collective engineering judgment of the team members, augmented by gravity load analyses with relevant structural members removed. The gravity load analyses primarily used a combination of linear response models and plastic analysis calculations. The NLFEA were used to a limited extent to validate the gravity load analyses. The projections do not involve rigorous progressive collapse computations for the entire building; they employed the same approach as used in the FEMA 439A investigation. Progressive collapse projections are reported in Chapter 4.

General conclusions are provided in Chapter 5.

1.4 Idealized Building

1.4.1 Basis for Building Selection

It was considered very important for the study to use a building that had been designed and constructed around the same time as the Murrah Building. This brings an important level of practicality to the study and makes it analogous to FEMA 439A. It was necessary from a security standpoint to protect the identity of the building. The existing building was selected and minor façade and layout modifications were made to alter its appearance. The modifications do not alter the structural system in any significant way.

In selecting the building, several general criteria were set:

- The focus would be on a structural steel building that relied on moment frames for lateral force resistance.

- Since the Murrah Building was a mid-rise structure, the steel building would be of similar height, between four and eight stories.

- The building would originally have been located in a region where seismic design and detailing would not have been required to satisfy the governing building code, again as was the case with the Murrah Building.

- The building must have been designed and constructed in the same era as the Murrah building, between 1960 and 1980.

1.4.2 Building Description

Given the selection criteria, the building chosen was a six story steel frame building constructed in 1970. The building was designed with minimal or no consideration for seismic forces due to its location in a region designated as low seismicity. Exterior elevations are shown in Figures 1-2 and 1-3. Figure 1-4 shows a typical floor plan. This building represents a common style of steel framed buildings built during the late 1960s to early 1970s throughout the United States.

The building is rectangular in plan, 180 feet long in the East-West direction by 150 feet wide in the North-South direction. The roof height is 83 feet above grade. Typical story height is 12 feet 6 inches, with the sixth story being slightly taller and the first story being 20 feet 6 inches tall. There is a 14 feet tall by 131 feet by 111 feet mechanical penthouse centered on the roof. There are two full basements below grade with floor-to-floor heights of 13 feet 6 inches and 14 feet, totaling 27 feet 6 inches below grade. Basement walls are 12-inch thick, two-story tall, reinforced concrete retaining walls. Figure 1-5 shows the foundation plan.

The exterior façade from the second floor to the roof consists of precast concrete panels with windows. The panels are 4 inches thick. Each individual panel is one story tall and extends an entire bay between columns. There are six window openings, with each panel being 6 feet tall by 3 feet 6 inches wide. Around each window are 6-inch tapered concrete fins that protrude out 2 feet from the panel face. Panels are reinforced with #3 or #4 bars spaced at 12 inches on center. Windows are inoperable, with ¼-inch annealed glass in aluminum window frames.

The exterior façade of the building is at the exterior of the structural framing except at the first story where it is inset one bay in each direction to create a plaza at ground level, although the structural system is continuous. Perimeter columns at the first story are encased with 2-inch marble panels. The stepped-in ground level exterior façade is a combination of full height marble panels and full height glass curtain walls.

Interior architectural finishes from the first basement level through the sixth floor level of the building consists of office spaces separated by movable partition walls. The main entrance and lobby are located along the west end of the first floor. The first basement also houses a mechanical room and storage space. The second (lower) basement serves as an underground parking garage and also contains storage space. The mechanical penthouse contains the majority of the mechanical equipment and the elevator lifts. There are two banks of elevators, two exit stairs, and two mechanical utility shafts, all located one bay west of the center of the building.

1.4.3 Structural System Description

The gravity load system typically consists of non-composite concrete on metal deck slabs set atop steel beams spaced at 10 ft on center. The steel beams frame into steel girders spaced at the column grid line spacing (30 feet on center). The girders frame into steel columns. All steel members are wide flange (WF) sections, with standard sizes from the Sixth Edition of the AISC *Manual of Steel Construction* (Reference 10). Beams are typically 18WF sections. Girders range from 16WF through 27WF sections. Columns are all 14WF sections. The metal deck is 3 inches thick, with 2-½ inches of concrete fill (5-½ inches total thickness) that is reinforced with welded-wire fabric.

The first floor and first basement floor have 5-inch cast-in-place concrete floors, reinforced with #4 reinforcement placed 12 inches on center each way. At those levels, the steel beams, girders, and columns are encased in concrete, however it is not known if the concrete is architectural, structural, fireproofing, or some other purpose. Connections of the first floor and basement girders to the columns are bolted shear tabs. Perimeter steel columns are based within concrete pilasters below the first floor framing. Therefore, the girders at the perimeter are framed into the steel columns through block-outs in the pilasters within the basement walls. The second basement floor is a 6-inch concrete slab-on-grade.

The foundation is a combination of the 12-inch basement retaining walls and spread footings. Interior columns bear on 25 feet square spread footings. Perimeter columns are embedded in pilasters cast monolithically with the basement wall. Under the pilasters are spread footings, 12 feet square at the corners and 16 feet square elsewhere. The wall is supported on a 2 feet wide continuous strip footing.

Vertical elements of the lateral force resisting system are moment frames. All girders are connected to the columns with full moment connections from the second floor to the roof, with the exception of the four corner columns on each floor, where moment connections are omitted at the girder to column web connections. Therefore, every frame in the structure is a moment frame. Elevations of typical exterior and interior longitudinal and transverse frames are shown in Figures 1-6 through 1-9. To distribute lateral stiffness evenly throughout the moment frames, the column orientations are alternated such that in the North-South (transverse) direction Lines A, C, E, and G have all the columns oriented with their weak axes in the plane of the frame and Lines B, D, and F have columns oriented with their strong axes in the plane of the frame. In the East-West (longitudinal) direction, Lines 1 and 6 have their columns' weak axes oriented in the plane of the frame (except for corner columns). Lines 2, 3, 4, and 5 have their columns' axes alternating between the strong and weak orientation.

Figure 1-10 shows a typical girder-to-column connection. The connection is representative of the standard Welded Unreinforced Flange (WUF) connections that performed poorly in the Northridge Earthquake (see FEMA 350 and FEMA 351). They were used extensively from the 1960s through the mid 1990s, because of their ease of fabrication and construction. Column splices (Figure 1-11) are located 1 foot 6 inches above floor levels and consist of partial-penetration welds along each of the flanges. Webs are not welded to the shear tab. Interior columns all have base plates embedded 12 inches below the slab-on-grade, with only two anchor rods anchoring them into the concrete footing pedestals. Perimeter columns are embedded in the pilasters within the concrete wall, as shown in Figure 1-12.

Concrete-filled metal decks serve as the horizontal elements (diaphragms), delivering inertial forces to the moment frames. The diaphragms are connected to the moment frames with ¾-inch diameter puddle welds at a 12-inch spacing. As-built drawings indicated that the sides of the metal deck panels were not fastened with side lap connections.

Based on the beam sizes in the moment frames, it appears that gravity (dead and live) loads governed the design, and not lateral (wind or seismic) loads. The beams are essentially the same size at all floor levels. Had wind or seismic loads governed, beam sizes would likely increase downward from the roof to the second floor. Also, the column splicing shown is a typical gravity load splice detail, in which minimal shear is assumed in the column, with corresponding flexural demands' being small enough to minimize the need for flanges to resist tension.

1.5 Limitations of the Study

There are four main issues that limit the applicability of this study. Each will be discussed in more detail in later sections of this report.

First, the study focuses solely on the issue of a steel moment frame system that is characteristic of those often found in older buildings and in buildings that are located in areas of low seismicity. Such moment frame systems are principally designed to resist gravity and wind loads; they have only minimal seismic energy dissipation capacity. Conclusions drawn from this study are not necessarily applicable to other structural systems, and they should not be considered to be prescriptive in any way.

Second, only the Murrah Building bombing scenario (a 4,000 lb TNT charge detonated at close range) is considered. It is a severe test and also serves as a "common denominator" for limited comparison to the FEMA 439A report conclusions. Other scenarios were beyond the scope of this study.

Third, the selected earthquake demands were based on a high-seismicity site, as discussed in section 1.3, resulting in special moment frame details. This limitation was imposed to keep the study scope at a manageable size. Because only one level of seismicity could be chosen, the most severe seismic environment was used to create an upper-bound data point for the study. The possibility of gaining ductility and toughness from designing intermediate moment frame details to accommodate a moderate seismic demand was not examined; that could be a point for future study.

Finally, the existing knowledge base regarding the blast response of structural steel elements is limited. Ongoing field tests that are sponsored jointly by DTRA and GSA are among the first such tests of structural steel that have been conducted. Those tests have been instrumental in validating the NLFEA that are described in Section 3.8. In addition to the DTRA-GSA testing for this project, the American Institute of Steel Construction (AISC) sponsored a field blast test of a steel column that was very similar to the ground story columns studied herein. The AISC testing, which is summarized in Section 3.9 and described in more detail in Appendix F, was very helpful in corroborating the analytical modeling that has been performed.

The lack of field testing of structural steel members contrasts the several decades of blast testing of reinforced concrete structures to support military programs. As a result, numerical models, particularly SDOF models, could not be validated as thoroughly for structural steel blast response as they have for reinforced concrete blast response.

Readers should be cautioned about making direct comparisons between this study and the study performed on the Murrah Building (FEMA 439A). While the study approach, loading, and analysis procedures are similar, there are differences in building configuration, member sizes, frame spacing, and building regularity. These differences have an impact on the overall behavior of the structures. Comparison of performance between the two buildings is not within the scope of this project.

1.6 Figures for Chapter 1

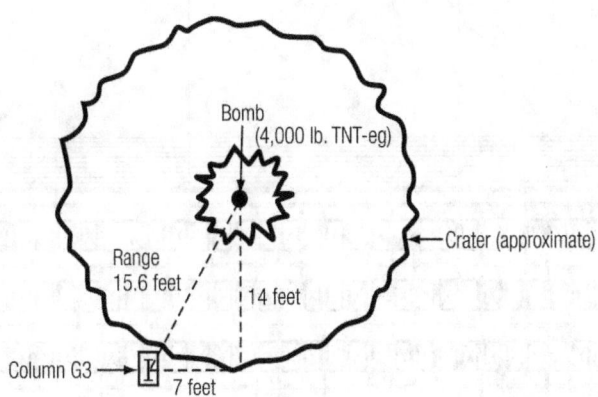

Figure 1-1. Bomb Location with Respect to Building

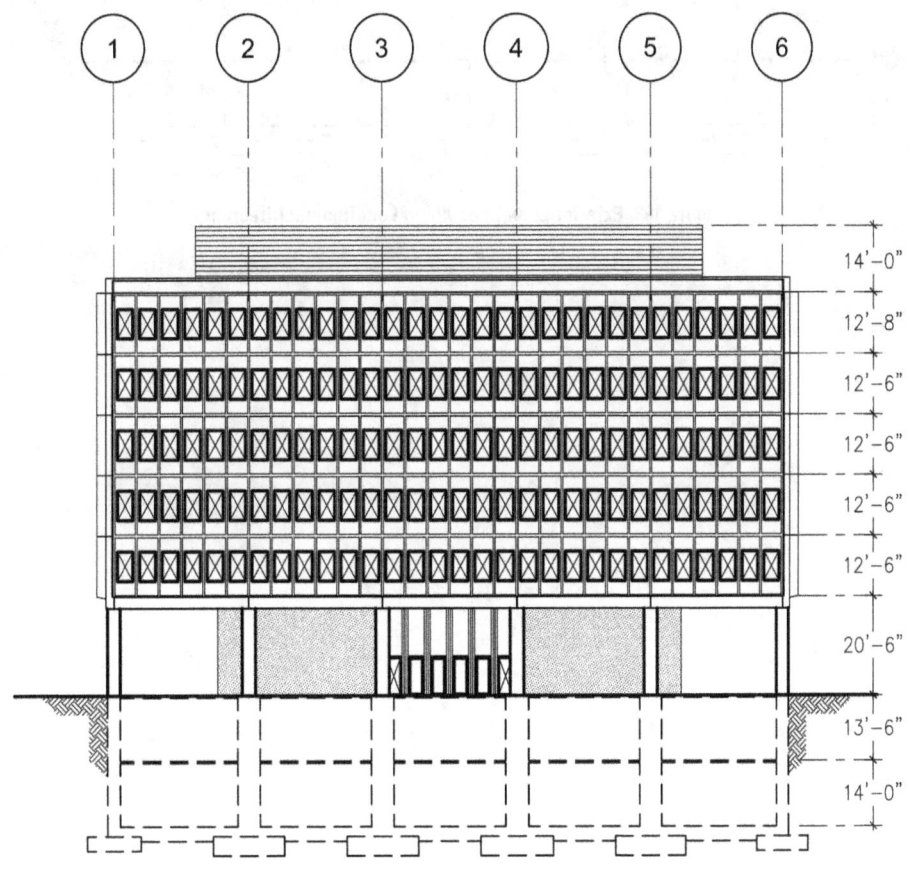

Figure 1-2. Exterior Elevations along Transverse Direction

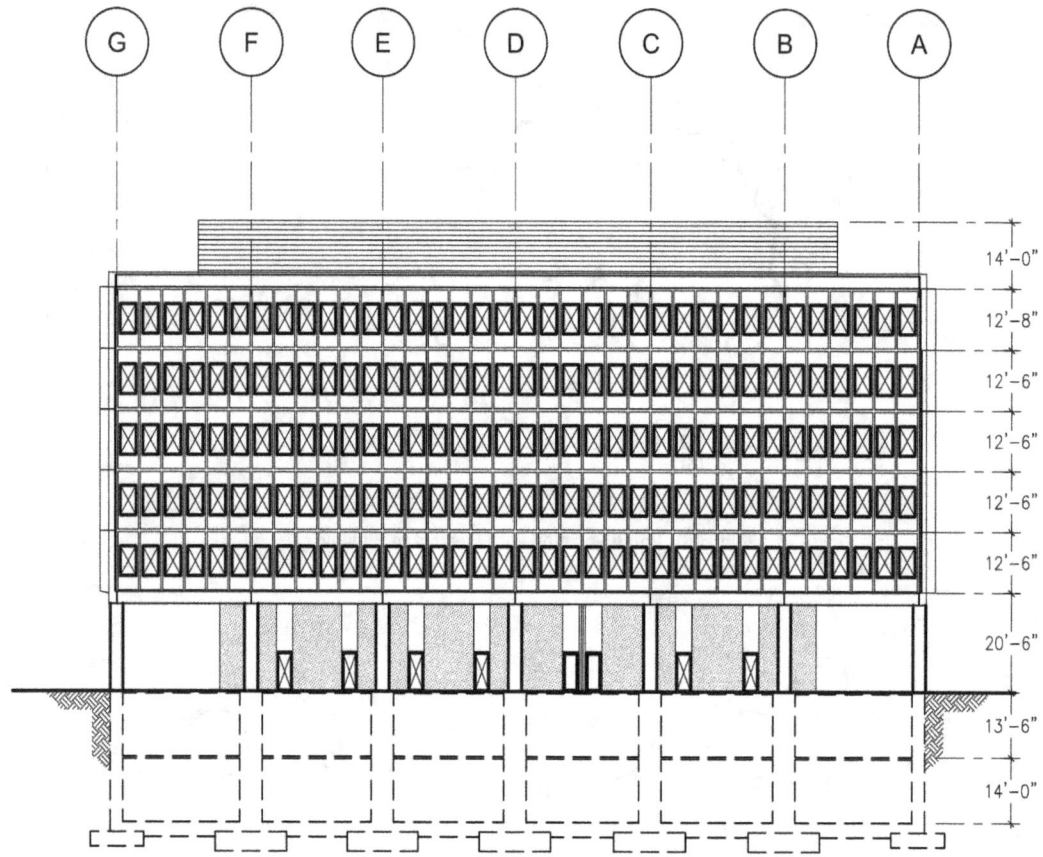

Figure 1-3. Exterior Elevations Along Longitudinal Direction

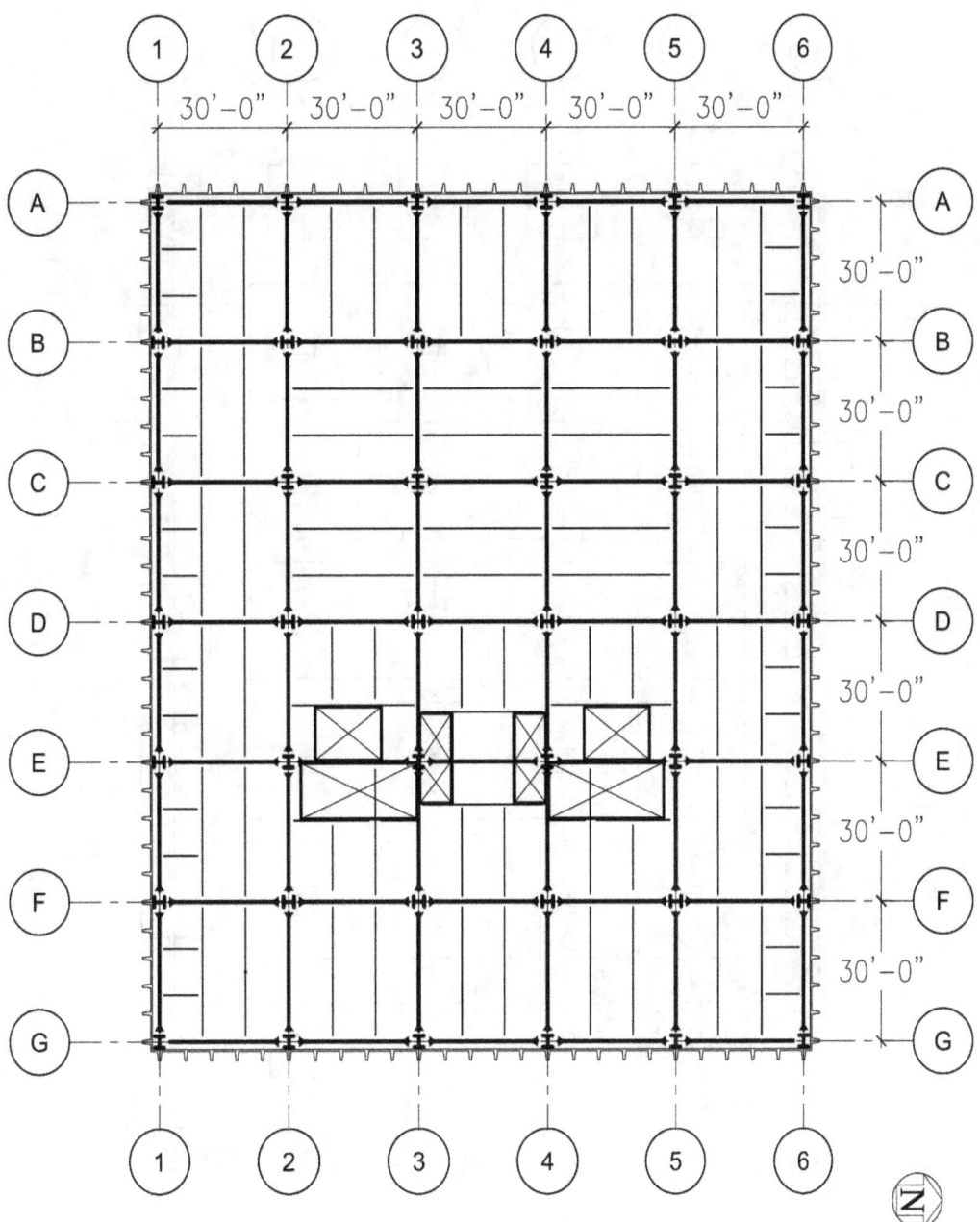

Figure 1-4. Building Floor Plan

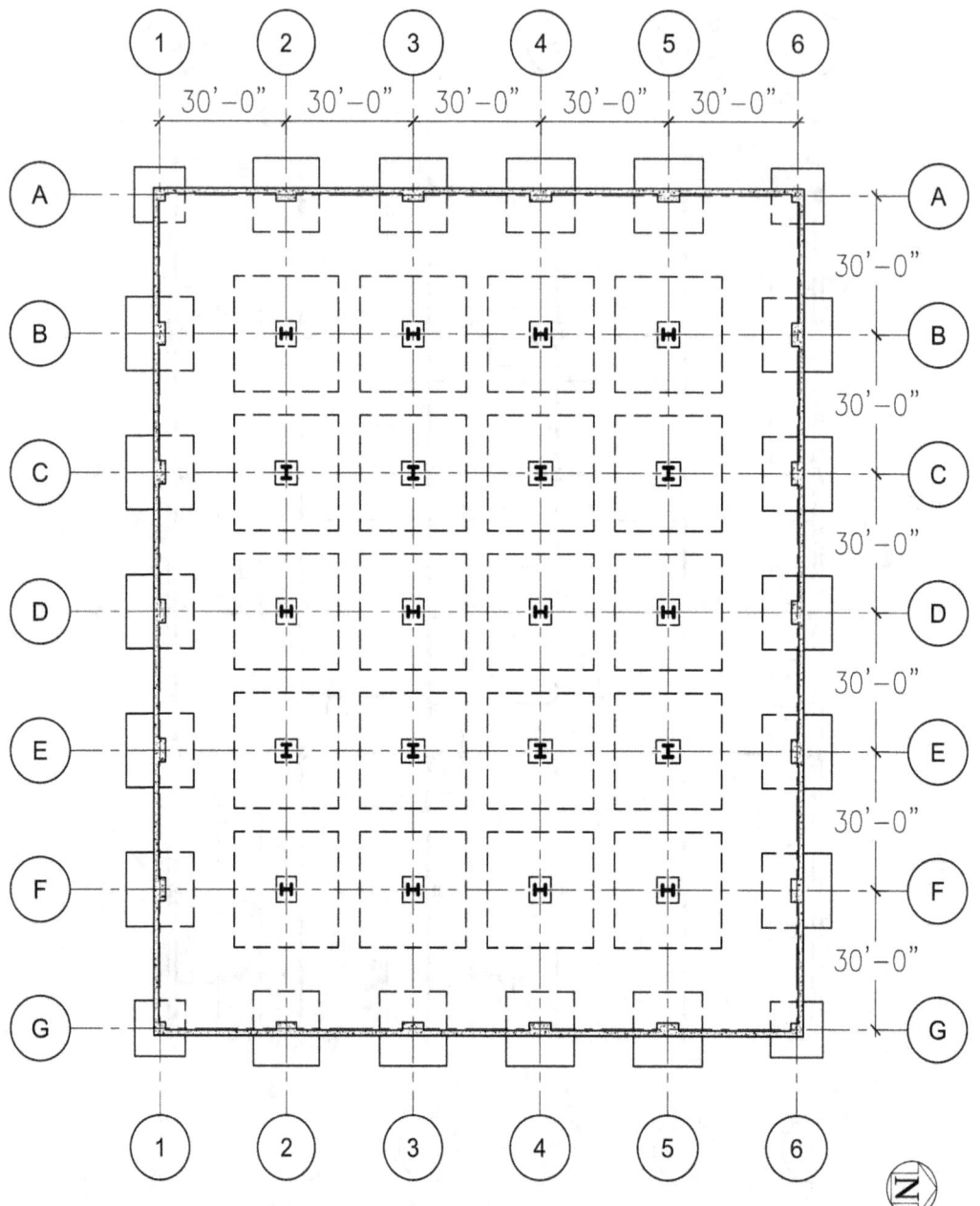

Figure 1-5. Foundation Plan

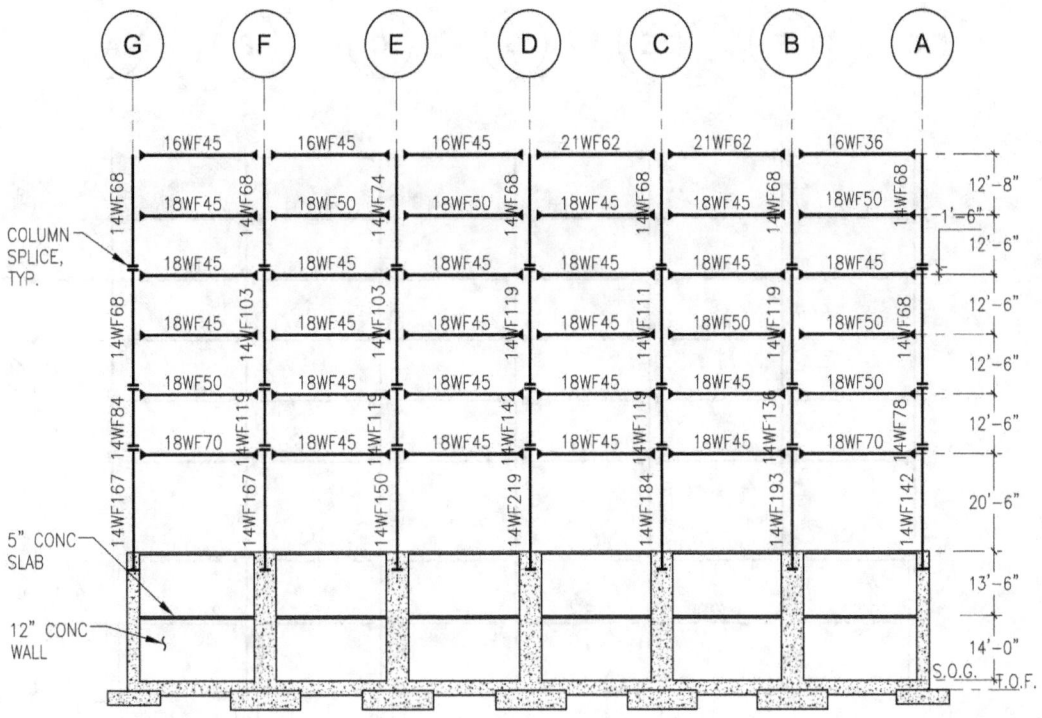

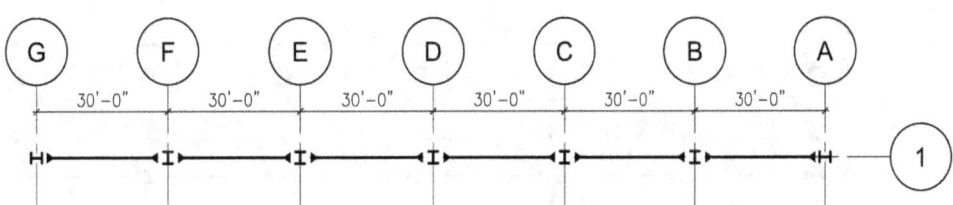

Figure 1-6. Typical Exterior and Interior Frame (Column Line 1)

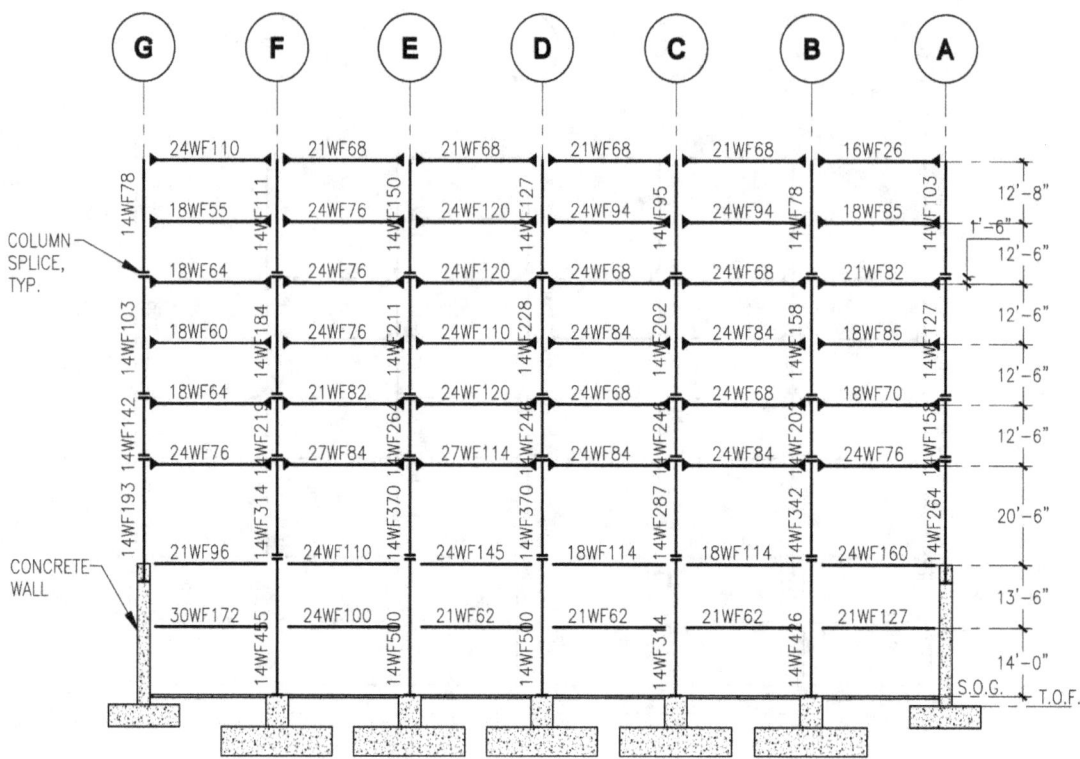

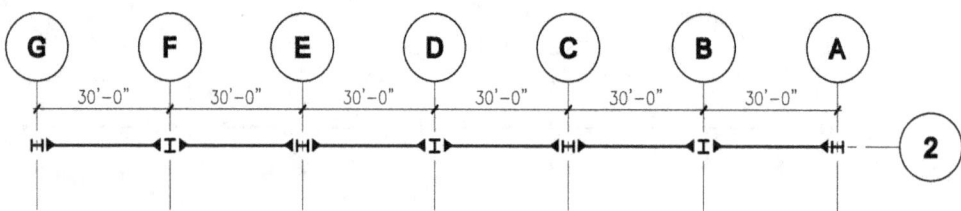

Figure 1-7. Typical Exterior and Interior Frame (Column Line 2)

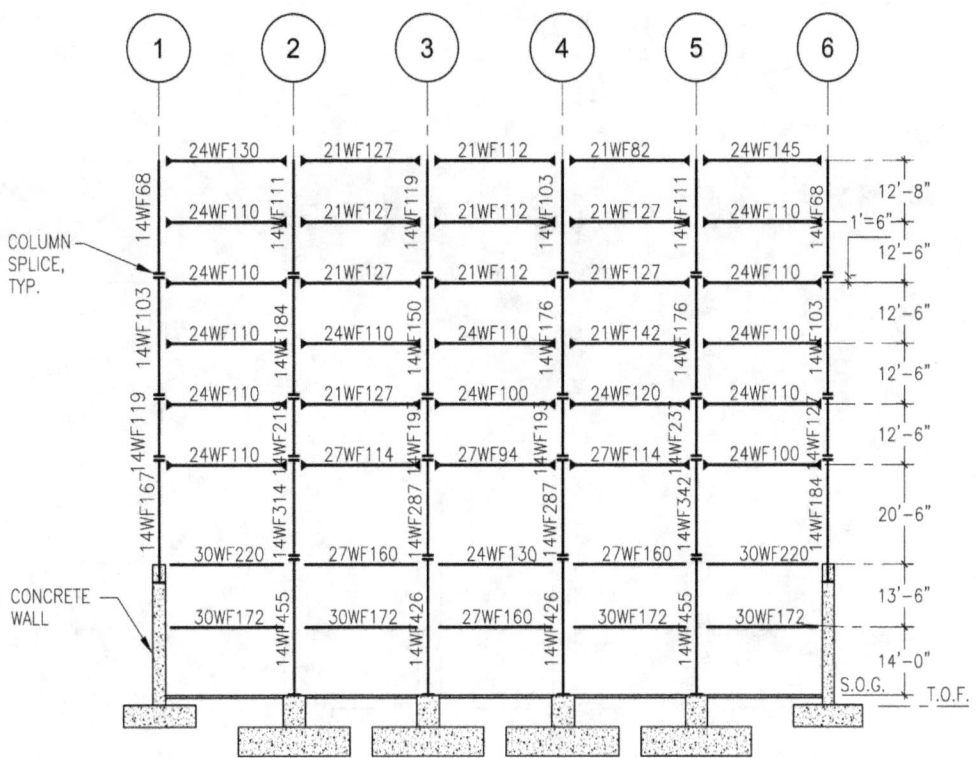

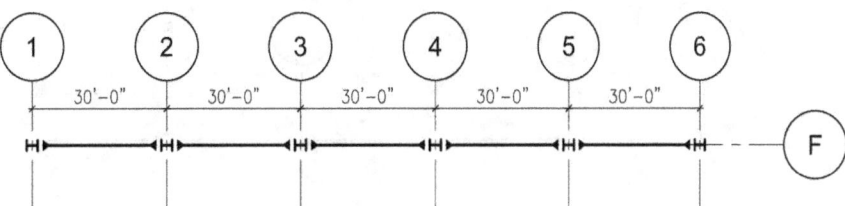

Figure 1-8. Typical Exterior and Interior Frame (Column Line F)

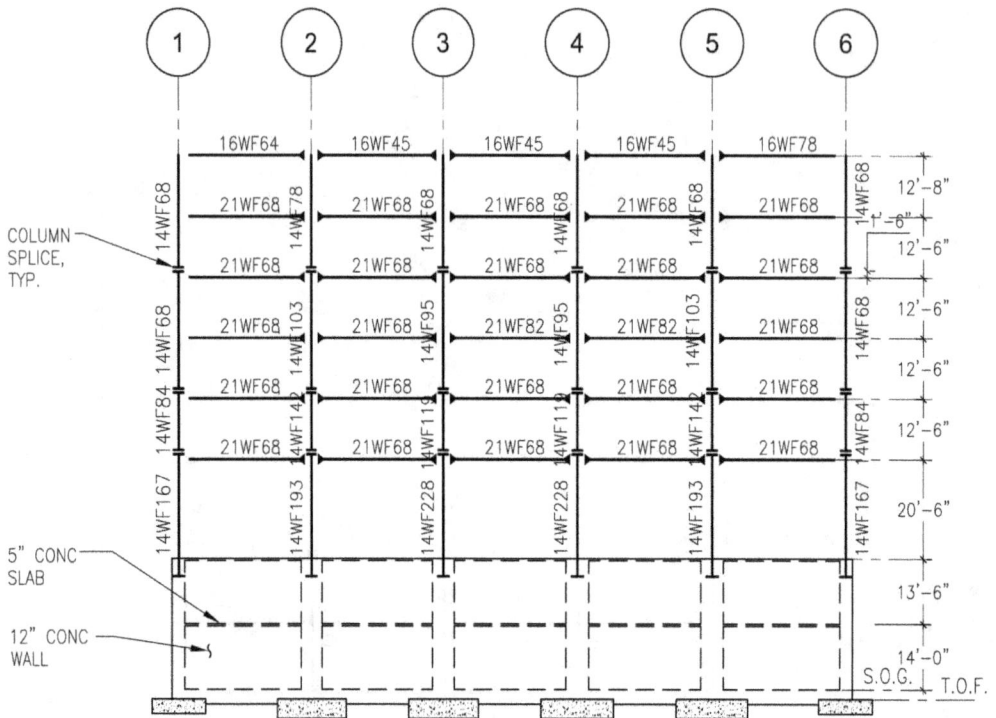

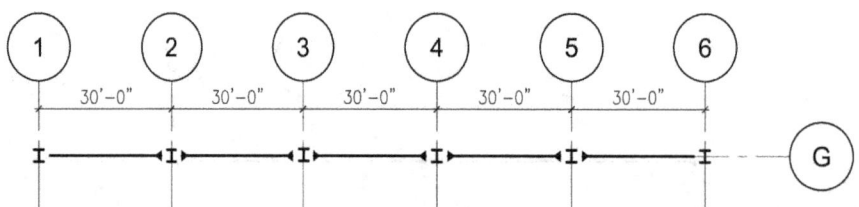

Figure 1-9. Typical Exterior and Interior Frame (Column Line G)

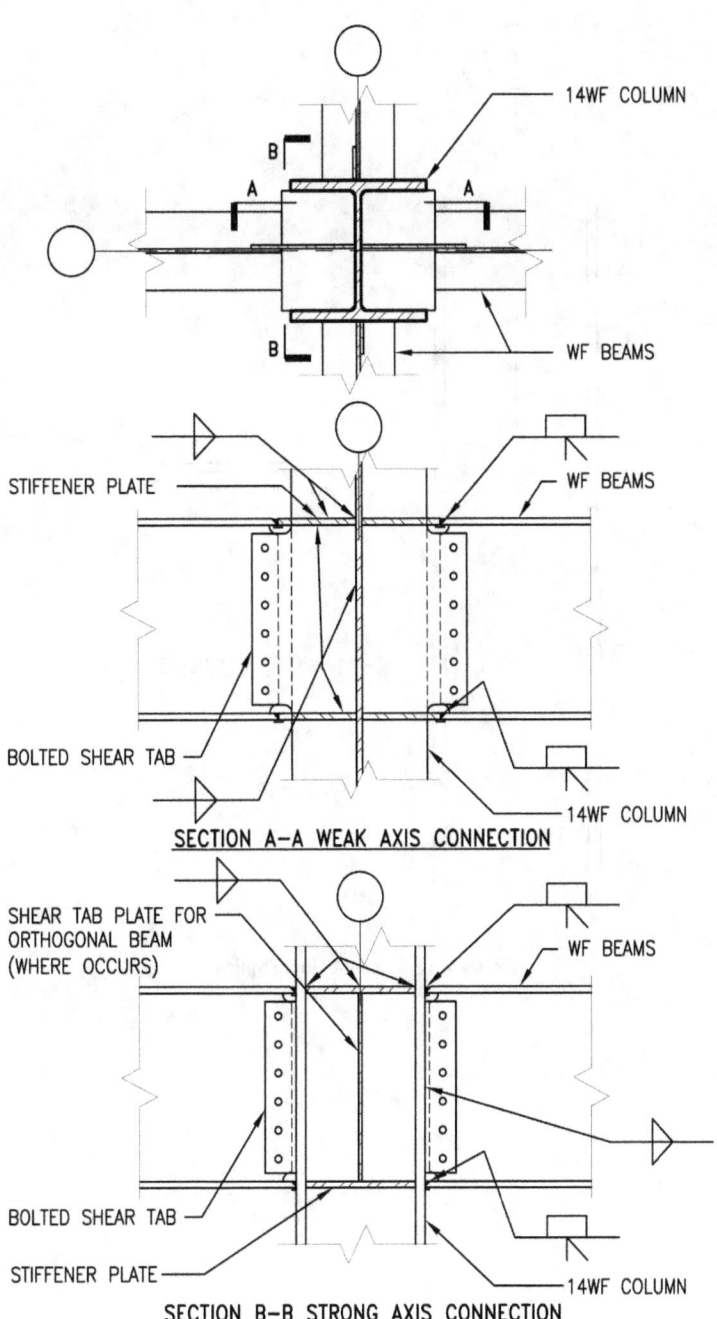

Figure 1-10 . Typical Beam-Column Connection

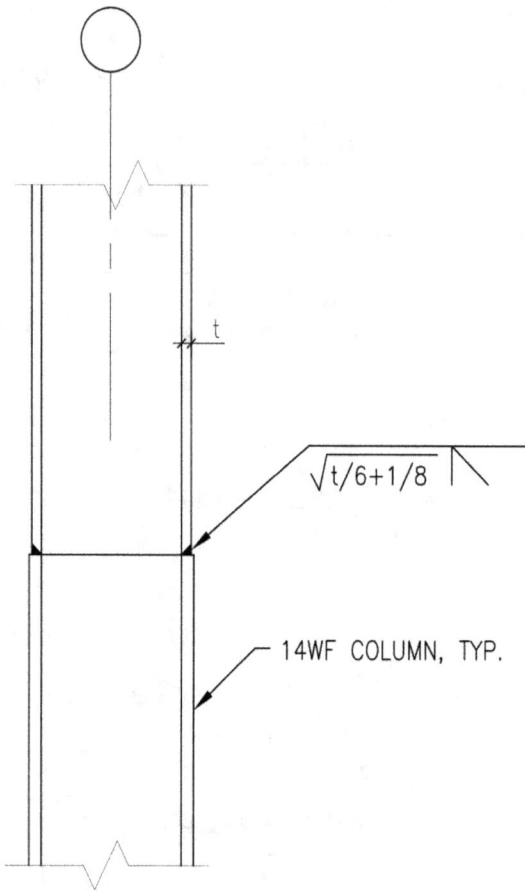

Figure 1-11. Typical Column Splice

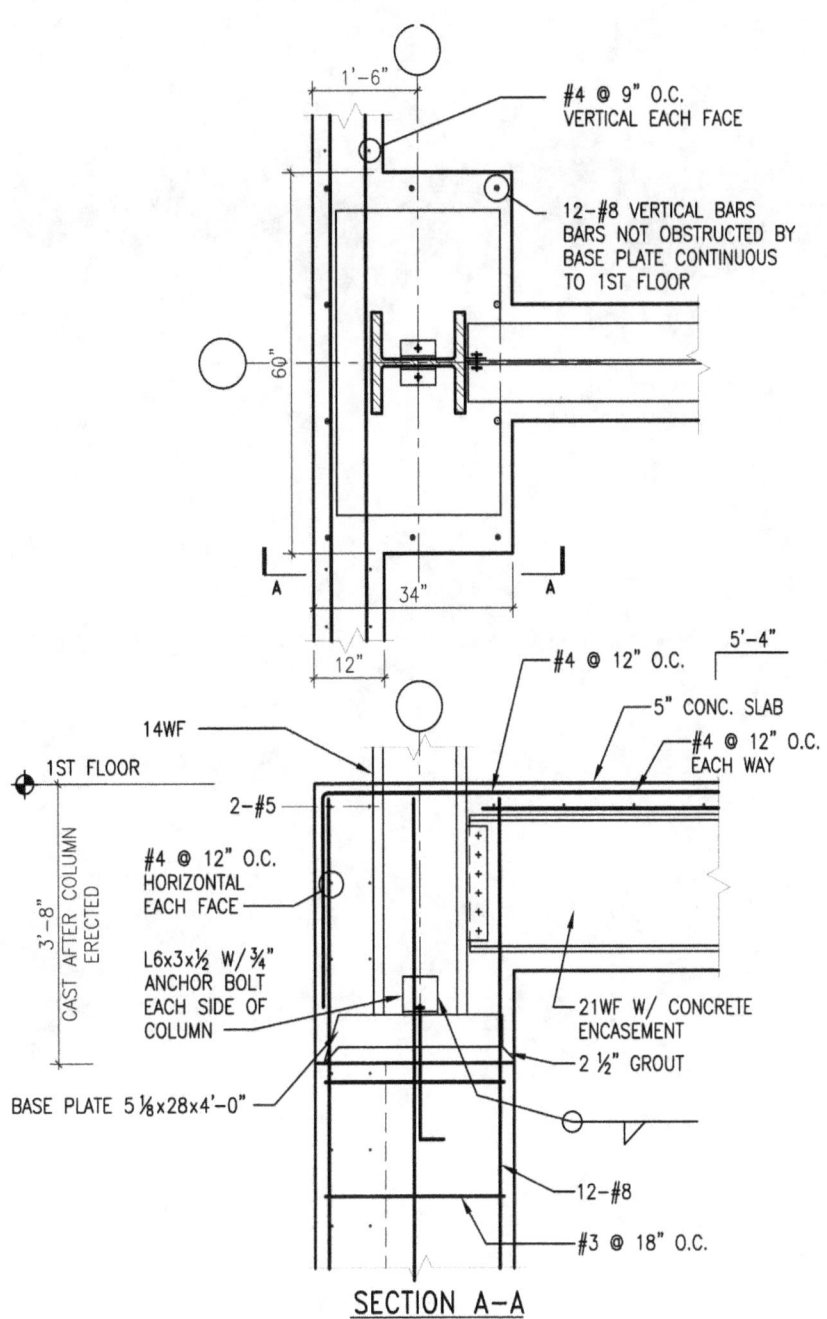

Figure 1-12. Typical Column Base

2   Seismic Evaluation and Strengthening

2.1   Seismic Evaluation

The building seismic evaluation was conducted using performance objectives and evaluation procedures outlined in ASCE 31-03, *Seismic Evaluation of Existing Buildings* (Reference 4). ASCE 31-03 is adapted from FEMA 310, *Handbook for the Seismic Evaluation of Buildings – A Prestandard* (Reference 5). FEMA 310 is in turn based on its predecessor, FEMA 178, *NEHRP Handbook for Seismic Evaluation of Existing Buildings* (Reference 41).

The building was evaluated for the basic seismic performance objective of *Life Safety*[1]. All three tiers of evaluation presented in ASCE 31-03 were performed: Tier 1 – Screening Phase, Tier 2 – Evaluation Phase, and Tier 3 – Detailed Evaluation Phase. The Tier 3 evaluation procedure is not prescribed in detail by ASCE 31-03, but instead allows the user to apply more rigorous analyses found in other documents. The Nonlinear Static Procedure (NSP) outlined in FEMA 356, *Prestandard and Commentary for the Seismic Rehabilitation of Buildings* (Reference 6), was utilized for Tier 3 evaluation in the current study.

As discussed in Chapter 1, seismic parameters for the study were based on a high seismicity location – IBC Seismic Design Category D (SDC D). To evaluate the steel building as if it had been located in an area of high seismicity, the building was 'virtually' relocated to the corner of 7th and Mission Streets in San Francisco, CA, instead of its original Seismic Design Category A site. Subsurface conditions at the new California site were similar to the building's original site, and the location is among the highest-seismicity locations in the United States.

Site-specific response spectra were developed for this location using the computer program *ST-Risk*[TM] (Risk Engineering, Boulder, CO, 2003). For the Tier 1 screening and Tier 2 evaluation specified by ASCE 31-03, spectral accelerations corresponding to two-thirds (2/3) of those of the BSE-2[2] event were used. For the Tier 3 evaluation specified

---

[1]   The Life Safety performance level is defined in ASCE 31-03 in terms of two primary criteria: (1) At least some margin against either partial or total structural collapse remains, and (2) Injuries may occur, but the overall risk of life-threatening injury as a result of structural damage is expected to be low.

[2]   The BSE-2, Basic Safety Earthquake-2 or Maximum Considered Earthquake (MCE), is defined as the ground shaking based on the combination of the ground shaking with a 2% probability of being exceeded within 50 years (approximately a 2,475 year return period) or with 150% of the median deterministically derived ground shaking at the given site. Source: FEMA 356.

by ASCE 31-03, and for subsequent seismic strengthening prescribed by FEMA 356, site-specific response spectra for the BSE-1[3] and BSE-2 events were used. The spectra are shown in Figure 2-1, which is a plot of the *ST-Risk* output.

Figure 2-1 shows that the BSE-2 short-period (0.2 sec) spectral acceleration is approximately 1.8 g, and the BSE-2 long-period (1.0 sec) spectral acceleration is approximately 1.5 g. These response spectra are for an estimated Site Class D/E. ASCE 31-03 Table 2-1 defines site seismicity in terms of thirds (2/3) of these values, or 1.2 g and 1.0 g for the short- and long-period spectral accelerations, respectively. ASCE 31-03 classifies "high" seismicity sites as those with spectral accelerations greater than or equal to 0.5 g and 0.2 g for the short- and long-period spectral accelerations, respectively.

The ASCE 31-03 assessment showed that, regardless of the level of evaluation rigor applied, the building would not meet the Life Safety performance level for a high-seismicity site; it would therefore need to be strengthened. The Tier 1 screening identified inadequate capacity of the lateral-force-resisting system for the seismic demand and inadequate detailing of moment frame elements to accommodate the deformations required of them for the ground motions used in the evaluation. Potential deficiencies that were identified include column splice and beam-column connection details, soft and weak story irregularities and excessive drift in the moment frames. The more refined Tier 2 evaluation confirmed Tier 1 findings, gave insights into the magnitudes of the deficiencies and eliminated some potential deficiencies. The Tier 3 evaluation showed more accurately what the demands were, relative to the building's capacity, as well as what measures would be required to strengthen the building to a Life Safety performance level.

The original steel building's design was in complete compliance with all circa-1970 requirements. It was in no way deficient in terms of 1960s to 1970s-era building codes for the given geographic location. Deficiencies found in the current ASCE 31-03 screening and evaluations are theoretical, based on an assumption of high seismicity and the application of much more stringent building codes than those applicable to seismic Zone 1 in the 1970s.

Complete screening and evaluation results are reported in Appendix A.

---

[3] The BSE-1, Basic Safety Earthquake-1, is defined as the lesser of ground shaking for an earthquake with a 10% probability of exceedence in 50 years (approximately a 475-year return period) or for an earthquake with 2/3 of the BSE-2 ground shaking. Source: FEMA 356.

2.2   Strengthening and Detailing for Improved Earthquake Performance

2.2.1 Strengthening Objectives

The seismic strengthening objectives were to eliminate brittle behavior of the original beam-column connections and column splices, and to eliminate instabilities reached prior to the BSE-2 target displacement. The intent was to accomplish these objectives with strengthening schemes that minimized architectural impact on the building. Three strengthening schemes were developed: an interior Buckling Restrained Braced Frame (BRBF) scheme, a connection upgrade scheme, and an exterior BRBF scheme that was combined with a hat truss. The intent of the first and third schemes was to stiffen and strengthen the building, which would decrease seismic demands in the beam-column connections and column splices to acceptable levels. The intent of the second scheme was to upgrade all beam-column connections and column splices with details capable of withstanding significant inelastic deformations without premature failure.

2.2.2 Interior BRBF Scheme

The first scheme utilized new Buckling Restrained Braced Frames (BRBFs) in four bays in each direction. This approach sought to mitigate seismic deficiencies by stiffening and strengthening the building. Figure 2-2 shows a typical floor plan with the locations of the new braced bays. Figures 2-3 and 2-4 show elevations of the transverse and longitudinal braced frame lines. Braced frame lines were located one column line inward from the exterior of the building. At the foundation level, braces were located to distribute overturning loads in a manner that would minimize the need for foundation strengthening and preclude the need to add piles. The foundation plan is shown in Figure 2-5.

The braces chosen were Buckling Restrained Braces (BRBs), shown in Figure 2-6. The BRBs were laid out in a chevron configuration within each bay, with brace strengths decreasing proportionally upward from the first story. At the top story, conventional braces consisting of hollow structural sections (HSS) were used instead of BRBs. This was done because the brace forces at that level were too low to justify the use of BRBs. Figure 2-7 shows a detail of the BRB connections. Columns of the braced frames were upgraded by first re-welding their splices (Figure 2-8) and then encasing them in concrete (Figure 2-9) to enhance their axial strength.

Existing connections of diaphragms to frame beams along those lines were inadequate and were upgraded by adding welded shear studs along the beams, as shown in Figure 2-10. Since those beams would then serve as collectors, shear tabs needed to be welded to beam webs to provide additional axial force transfer, and bottom flanges required bracing (Figure 2-11). For the second floor beams within the braced frames, WT sections were welded to bottom flanges, as shown in Figure 2-10, to upgrade their capacities to transmit forces into the first story braces.

To distribute foundation loads, transfer trusses were added below the first floor, and new grade beams were added between the braced frame columns, below grade. Column connections into the foundation were also upgraded to resist uplift forces generated by the frames. Figure 2-12 shows a detail of the connection of a column and brace gusset plate to the new grade beam.

2.2.3 Connection Upgrade Scheme

The intent of this scheme was to strengthen beam-column connections and column splices to provide the moment frames with required ductility. This scheme is shown in plan view in Figure 2-13. Beam-column connections were upgraded to the Welded Bottom Haunch (WBH) configuration outlined in FEMA 351 and shown in Figure 2-14. All moment frame beams required bottom flange bracing (Figure 2-11). Column splices were upgraded by back-gouging existing partial penetration flange welds and re-welding the splices with complete joint penetration welds, as shown in Figure 2-8. Webs were also welded together with complete joint penetration welds.

2.2.4 Exterior BRBF Scheme with Hat Truss

The final strengthening scheme deviated from the original directive slightly by including explicit consideration for progressive collapse mitigation within the seismic upgrade. This is referred to as a "Smart" scheme. In this scheme, the Buckling Restrained Braced Frames (BRBFs) were placed along the perimeter frames. Figure 2-15 shows a plan of this scheme. Transverse and longitudinal elevations are shown in Figures 2-16 and 2-17, respectively. Using this approach, most perimeter columns have braces attached to them, so if a column is lost, the braces could form a truss to support the columns above. However, since eight perimeter columns did not have truss support, a "hat truss" was created along the sixth story perimeter so that all columns were protected against progressive collapse. This was viewed as the most economical approach to supporting the eight remaining perimeter columns.

The Exterior BRBF scheme is similar to the Interior BRBF scheme except for braced frame locations. Brace configurations and sizes remained the same. Beams again required shear studs and bottom flange bracing to be added, and the second floor beams within the braced frames required WT sections to be welded to their bottom flanges. All braced frame columns required upgrading of splices and concrete encasement. The major difference between this scheme and the interior scheme is the foundation con-nection configuration. Figure 2-18 shows the connection of an existing column into the basement wall.

2.3   Application of Seismic Detailing to Original Design

One of the conclusions of FEMA 277 was that if the exterior frames of the Murrah Building had been detailed using the provisions for special reinforced concrete moment frames found in ACI 318-02 (Reference 43), Chapter 21, much of the damage caused by the blast and the ensuing progressive collapse could have been prevented. In reinforced concrete, seismic detailing is the process of designating the amounts, lengths, bends, and locations of steel reinforcement. The results of FEMA 439A showed this conclusion to be valid for a reinforced concrete frame. It was decided to test this hypothesis for a steel building detailed using the provisions for special steel moment frames found in AISC 341 and supplemented by the recommendations in FEMA 350. For structural steel, seis-mic detailing is the process of designating the section sizes, connection configurations and locations, and stability bracing for the steel frame.

To evaluate this hypothesis, the transverse and longitudinal frames of the original build-ing were re-detailed to make them comply with the special moment frame detailing pro-visions of AISC 341, supplemented by the recommendations in FEMA 350. This fourth scheme did not constitute a "seismic re-design" to meet specific force or displacement requirements associated with a specific seismic hazard. Figure 2-19 provides a plan of the seismically re-detailed frame showing the revised column orientations. Figures 2-20 and 2-21 show the transverse and longitudinal frames after being re-detailed.

For the re-detailing, a *prequalified* (see Section 3.4 of FEMA 350) beam-column con-nection was selected. The Reduced Beam Section (RBS) was chosen and is shown in Figure 2-22. Most beam sections remained unchanged, with the exception of a few along the longitudinal frames whose sizes had to be increased slightly to meet seismic compactness requirements. Column sizes were all increased so that panel zones would remain elastic or yield simultaneously with the beams. Because the prequalified connec-tion is required to have the inelasticity in the panel zone be consistent with that in the tested connection, it is sometimes common practice to increase column size instead of adding doubler plates to upgrade the strength of the panel zone. Larger column sec-tions are more economical than doubler plates because of the large cost associated with fabricating and installing the doubler plates. Column splices were moved to 4 feet above floor levels, and the number of splices used over the height of the framing was de-creased. The splices were made with full penetration welds. Figure 2-23 shows the new column splice detail.

The re-detailed scheme was then subjected to the same blast scenario as the original building and the three upgraded schemes, to determine what improvement could be gained from simply re-detailing the perimeter frames. The results of the blast analysis and the post collapse evaluation are presented in Chapters 3 and 4 respectively.

2.4   Estimated Costs of Upgrade Schemes

As with the Murrah Building Study (FEMA 439A), estimated costs were determined for the seismic upgrades and also for the re-detailed perimeter frame. The costs were determined using 2006 construction prices and based on the building being located in San Francisco. This is different from the Murrah building report, in which the construction costs were based on Oklahoma City using 2003 construction costs, so direct comparisons cannot be made without scaling factors that account for location and inflation of the costs. The upgrade costs are summarized in Table 2-1, and the detailed estimate is provided in Appendix E.

The costs for the seismic upgrade only vary by 20%, with the Exterior BRBF scheme being the least expensive at $7,515,000 and the Connection Upgrade being the most expensive at $9,146,000. The Interior BRBF scheme was estimated to cost $8,538,000. The reason for the Exterior BRBF scheme being less expensive than the Interior BRBF scheme is that even though the exterior scheme requires costly removal of the cladding panels, the Interior BRBF scheme requires substantial foundation work, which must be accomplished inside the structure. The cost of the connection upgrade scheme is highest because every connection must be upgraded, requiring demolition and reconstruction of the floor slab and interior finishes around every column. The cost of the Exterior BRBF with the "Hat Truss" scheme was $7,893,000. Adding the hat truss only represents an increase of 5% of the total construction cost.

When determining which seismic upgrade would be selected, cost is not the only factor that would come into play. While the Connection Upgrade is the most expensive, it is not necessarily an unappealing scheme because it preserves the functional flexibility of the floor plans. If the structure were to undergo a multi-hazard upgrade the windows would most likely be upgraded, which would make the Exterior BRBF scheme potentially more appealing. The most appealing aspect of the Interior BRBF scheme is that there are no braces obstructing the window openings, like the Exterior BRBF scheme. The tradeoff between the Interior BRBF scheme and the Connection Upgrade scheme is the loss of some functionality for better seismic performance.

In addition to estimating the costs of the seismic upgrade, the cost implications of re-detailing the original frame were determined. First, the cost of the original building if it were to be built at the time of this study (2006) was estimated to be $43,200,000. The cost of the perimeter frame as detailed on the original drawings and the cost of re-detailed perimeter frame were estimated to be $919,000 and $1,717,000, respectively. The increase of approximately $798,000 represents an increase of 1.8% of the total construction cost.

Table 2-1. Estimated 2006 Construction Costs (in San Francisco) for Seismic Strengthening Schemes

Strengthening Scheme	Estimated 2006 Construction Cost
Interior BRBF	$8,538,000
Connection Upgrade	$9,146,000
Exterior BRBF	$7,515,000
Exterior BRBF with "Hat Truss"	$7,893,000
Re-Detailing Costs	**Estimated 2006 Construction Cost**
Original Perimeter Frames	$919,000
Perimeter Frames Detailed As SMRFs	$1,717,000
Estimated Total Original Building Cost	$43,200,000
Increase in Total Building Cost	1.8%

2.5   Figures for Chapter 2

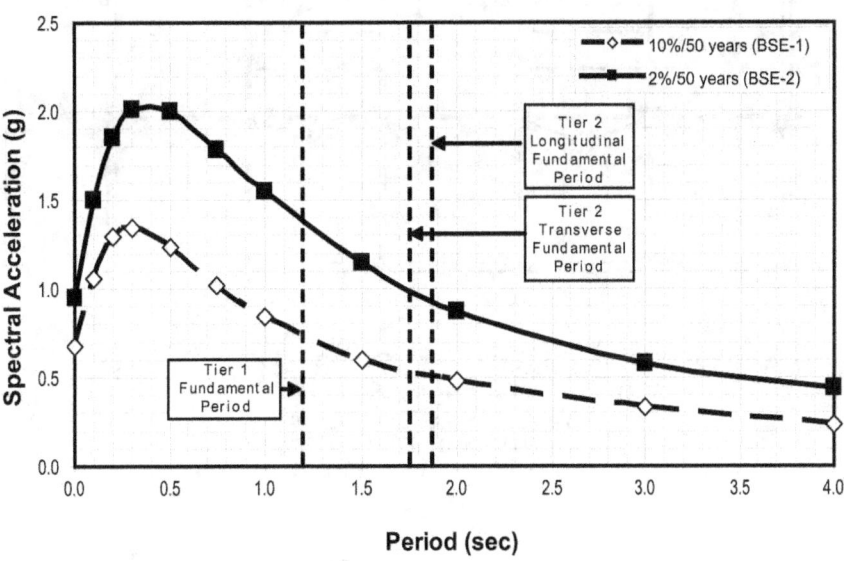

Figure 2-1. Site-Specific Response Spectra Derived Using *ST-Risk*

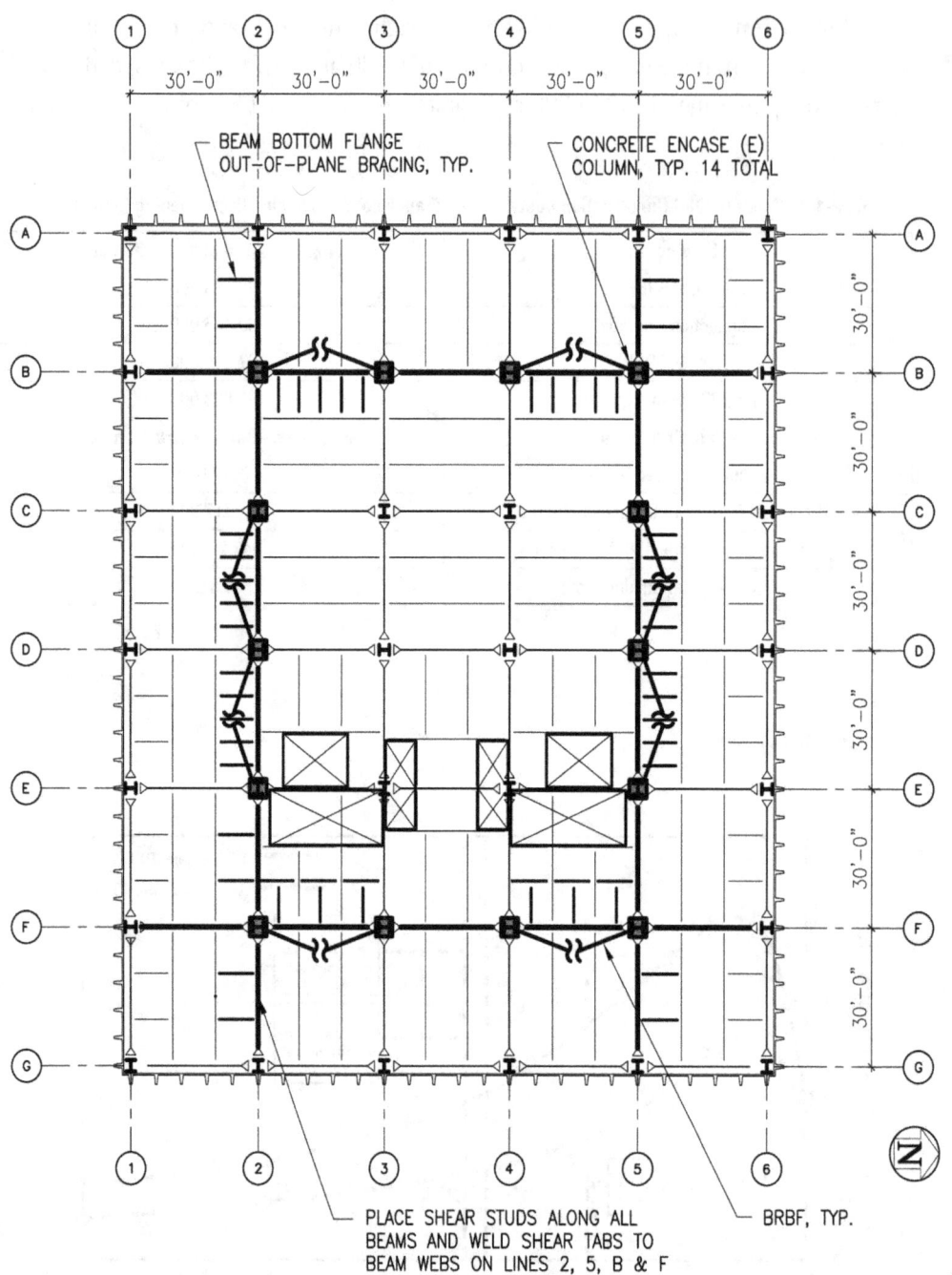

Figure 2-2. Interior BRBF Plan

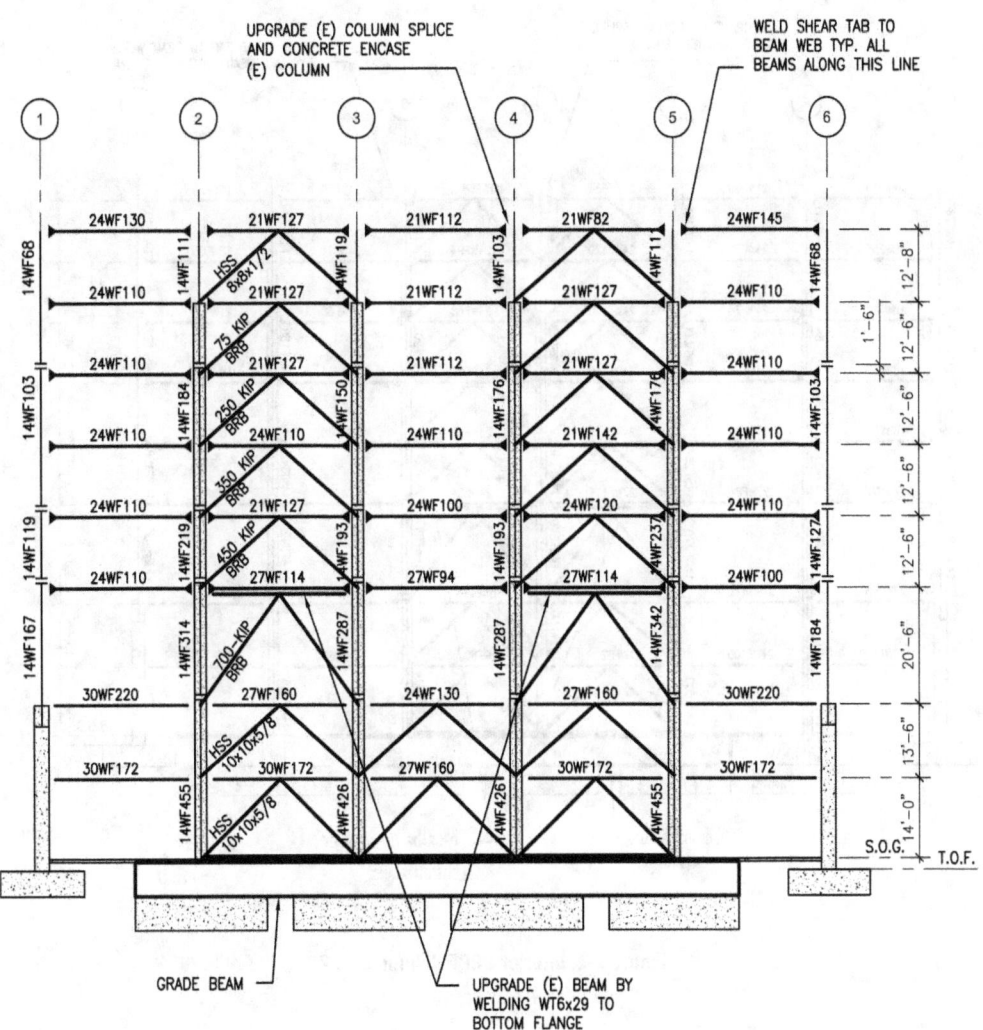

Figure 2-3. Interior BRBF Column Line F

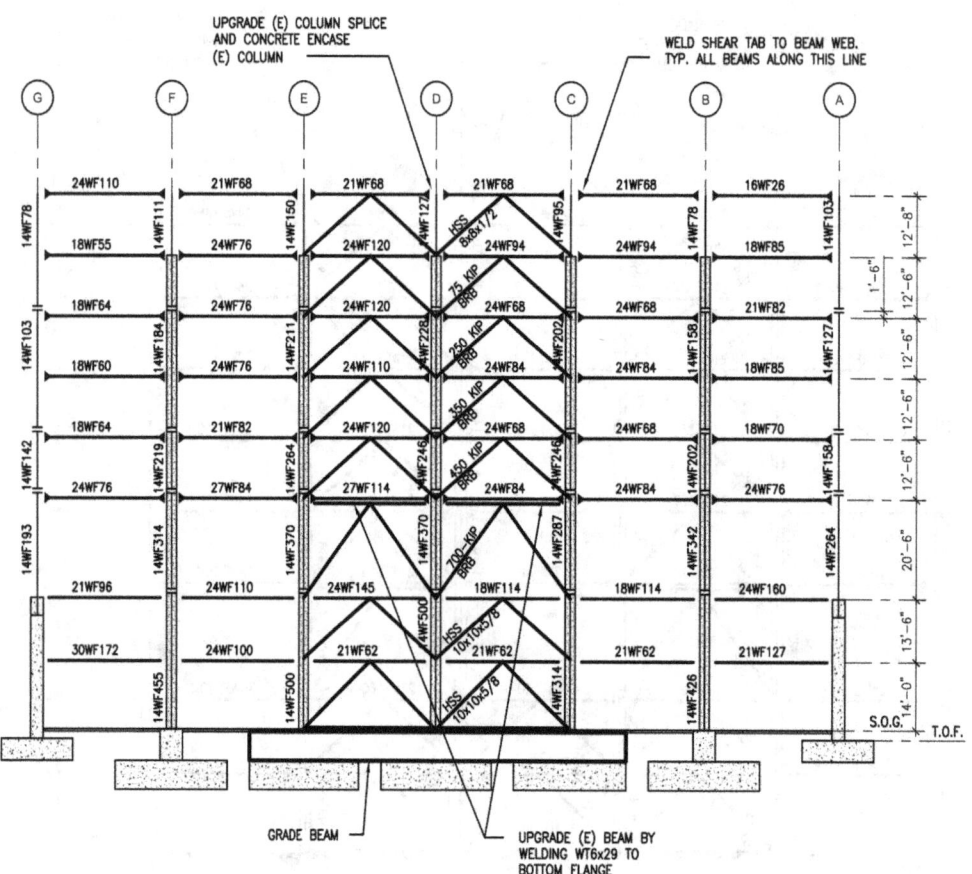

Figure 2-4. Interior BRBF Column Line 2

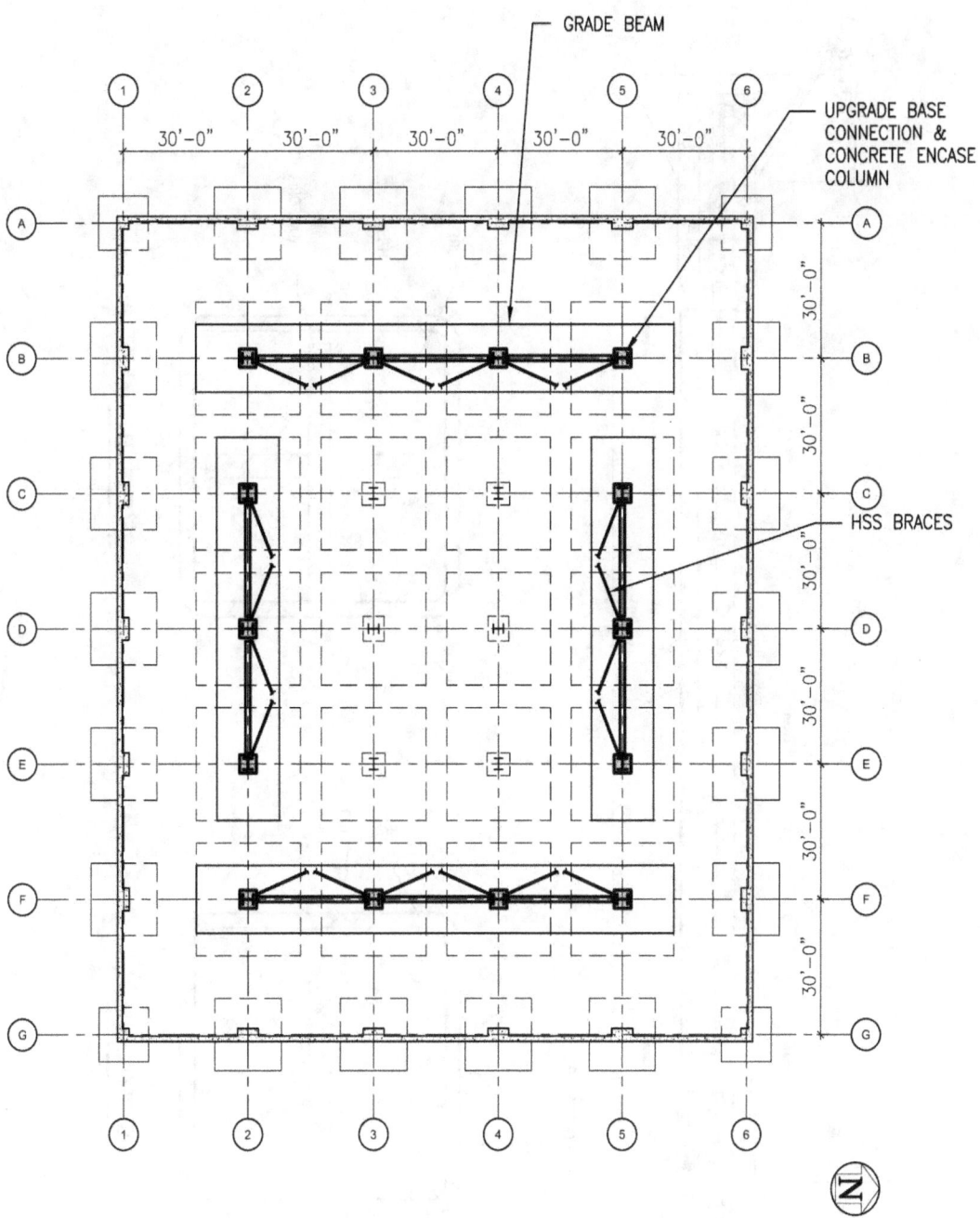

Figure 2-5. Interior BRBF Foundation Plan

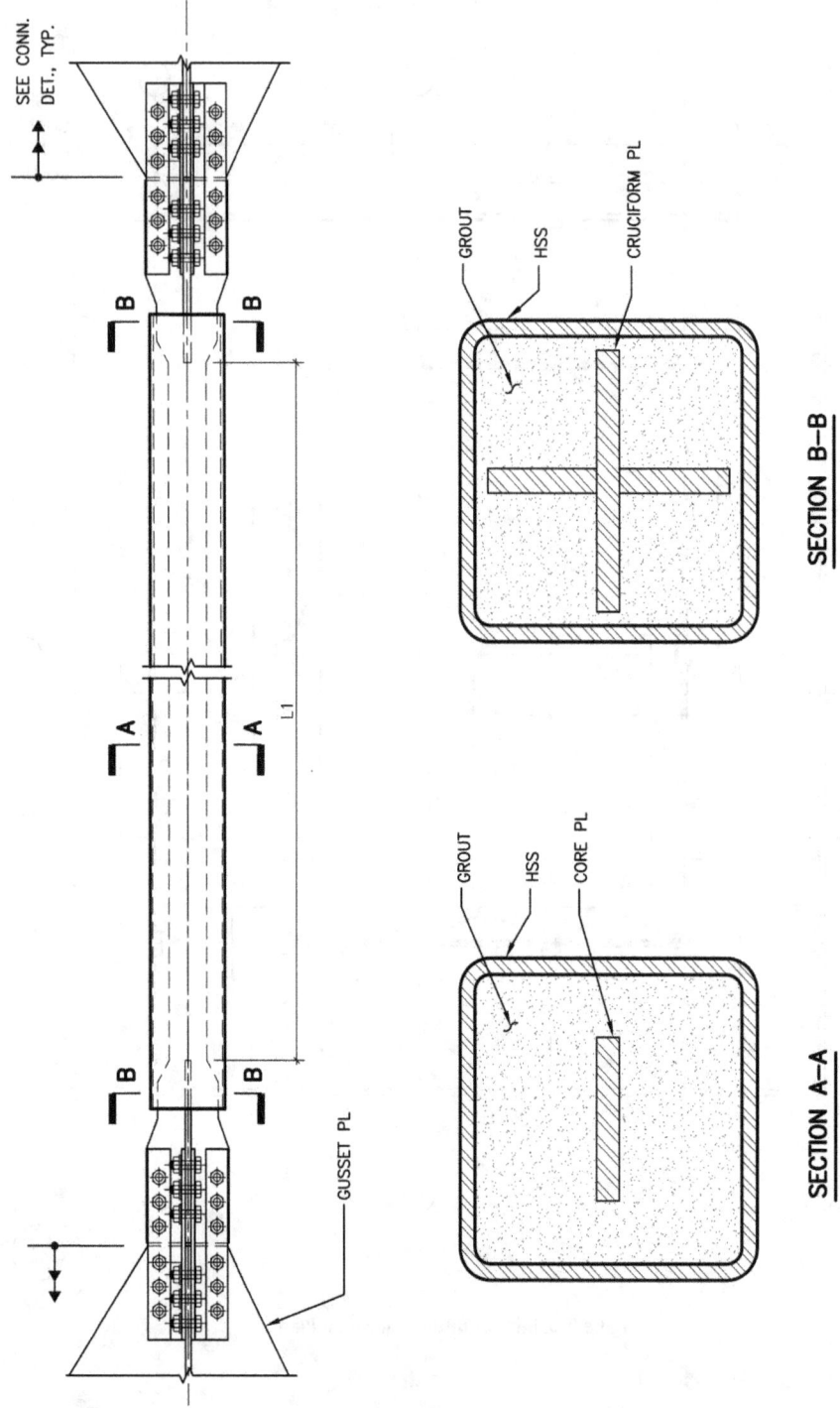

Figure 2-6. Bucking Restrained Brace (BRB)

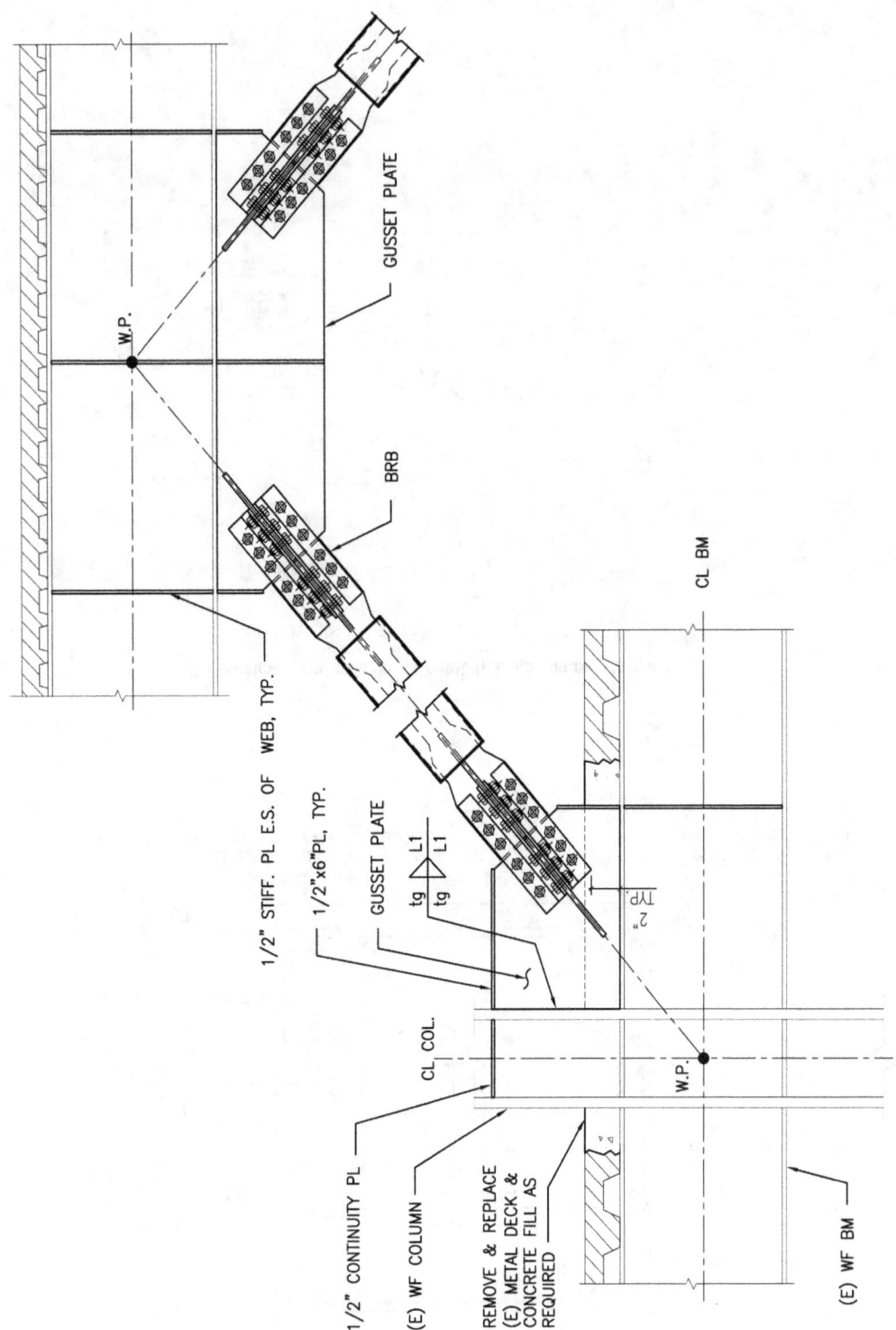

Figure 2-7. Detail of BRB Connections

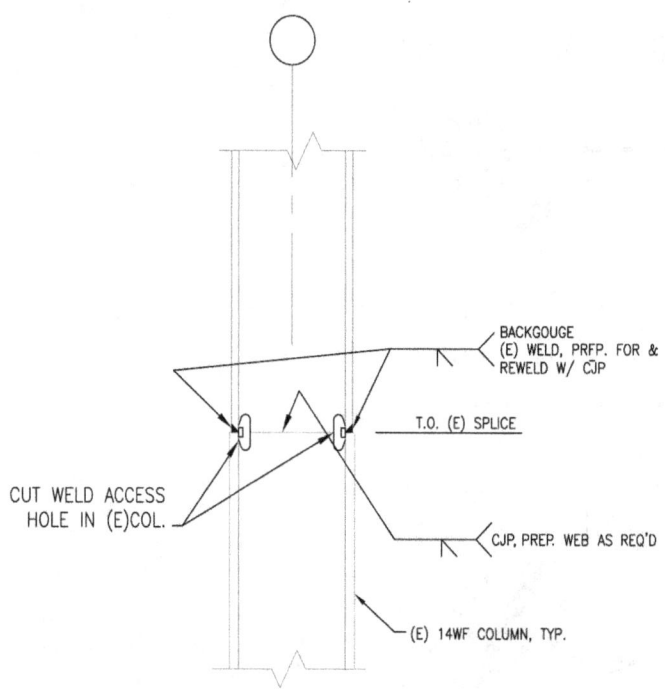

Figure 2-8. Column Splice Upgrade for Interior BRBF Scheme

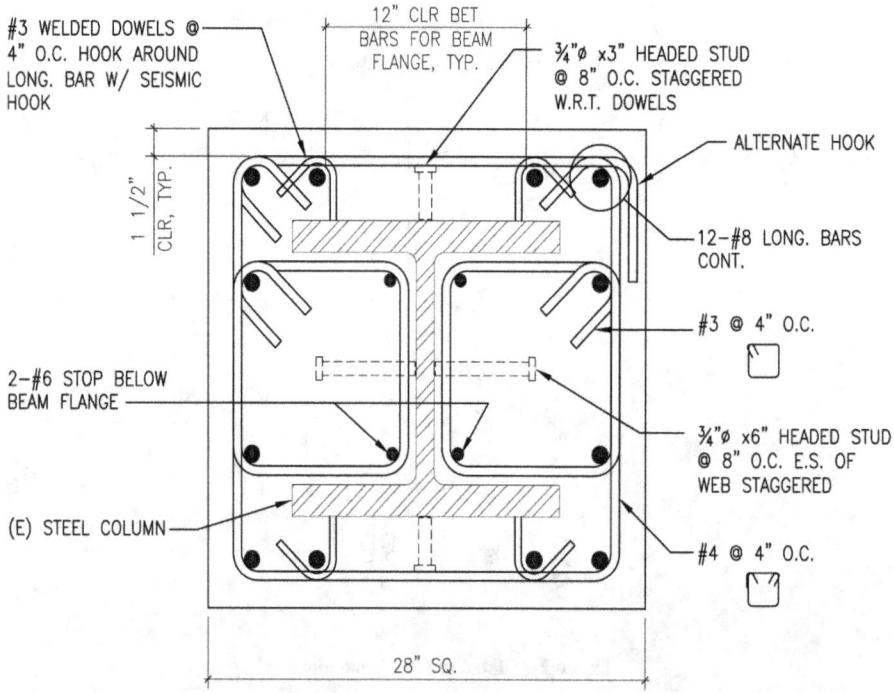

Figure 2-9. Column Concrete Encasement for Interior BRBF Scheme

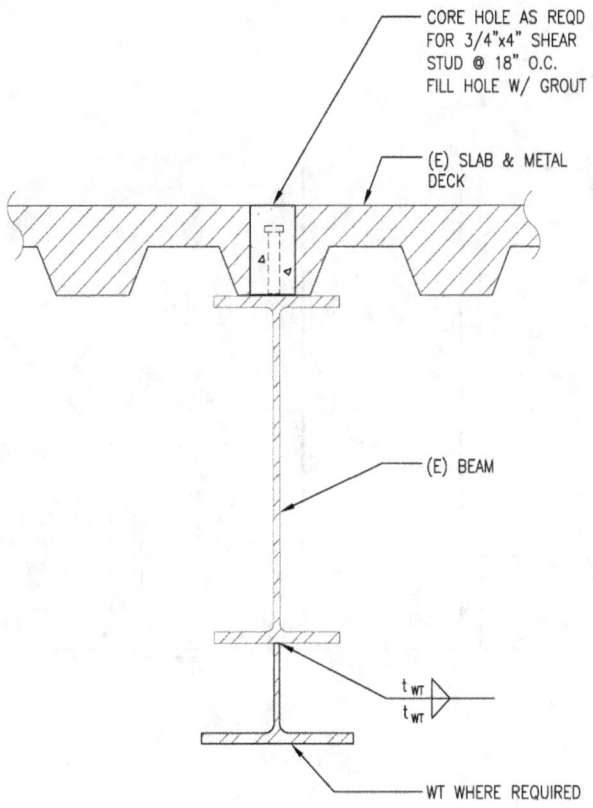

CORE HOLE AS REQD
FOR 3/4"x4" SHEAR
STUD @ 18" O.C.
FILL HOLE W/ GROUT

(E) SLAB & METAL
DECK

(E) BEAM

t_{WT}
t_{WT}

WT WHERE REQUIRED

Figure 2-10. Beam Upgrade for Interior BRBF Scheme

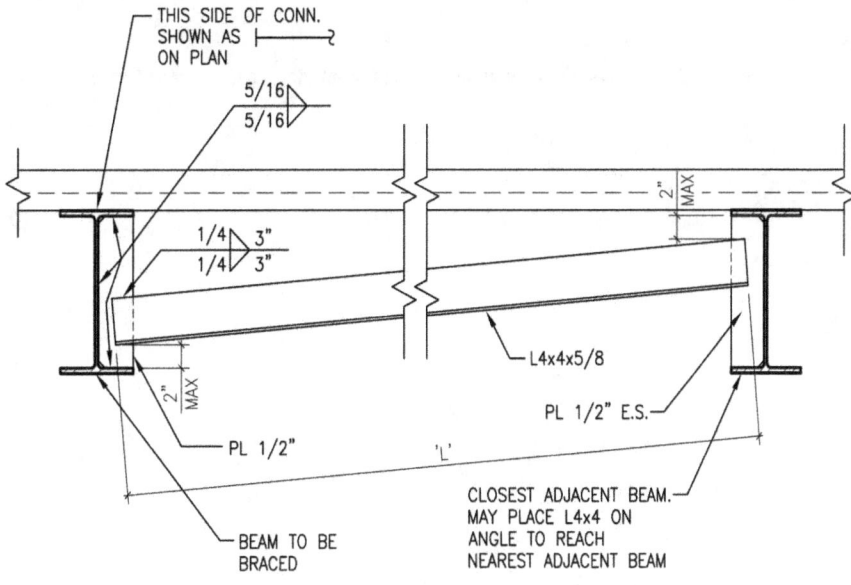

THIS SIDE OF CONN.
SHOWN AS
ON PLAN

5/16
5/16

2" MAX

1/4 3"
1/4 3"

L4x4x5/8

2" MAX

PL 1/2" E.S.

PL 1/2"

'L'

BEAM TO BE
BRACED

CLOSEST ADJACENT BEAM.
MAY PLACE L4x4 ON
ANGLE TO REACH
NEAREST ADJACENT BEAM

Figure 2-11. Beam Brace for Interior BRBF Scheme

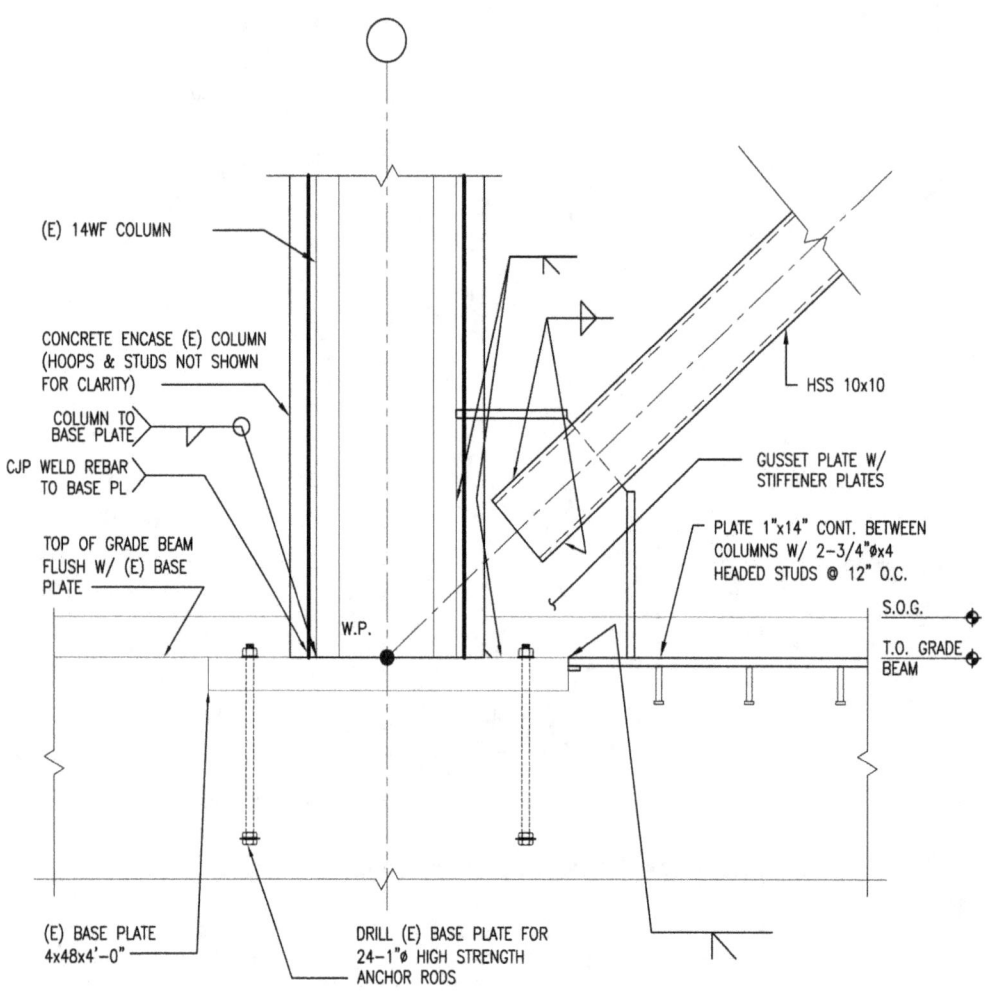

Figure 2-12. Connection of Column and Brace Gusset Plate to New Grade Beam

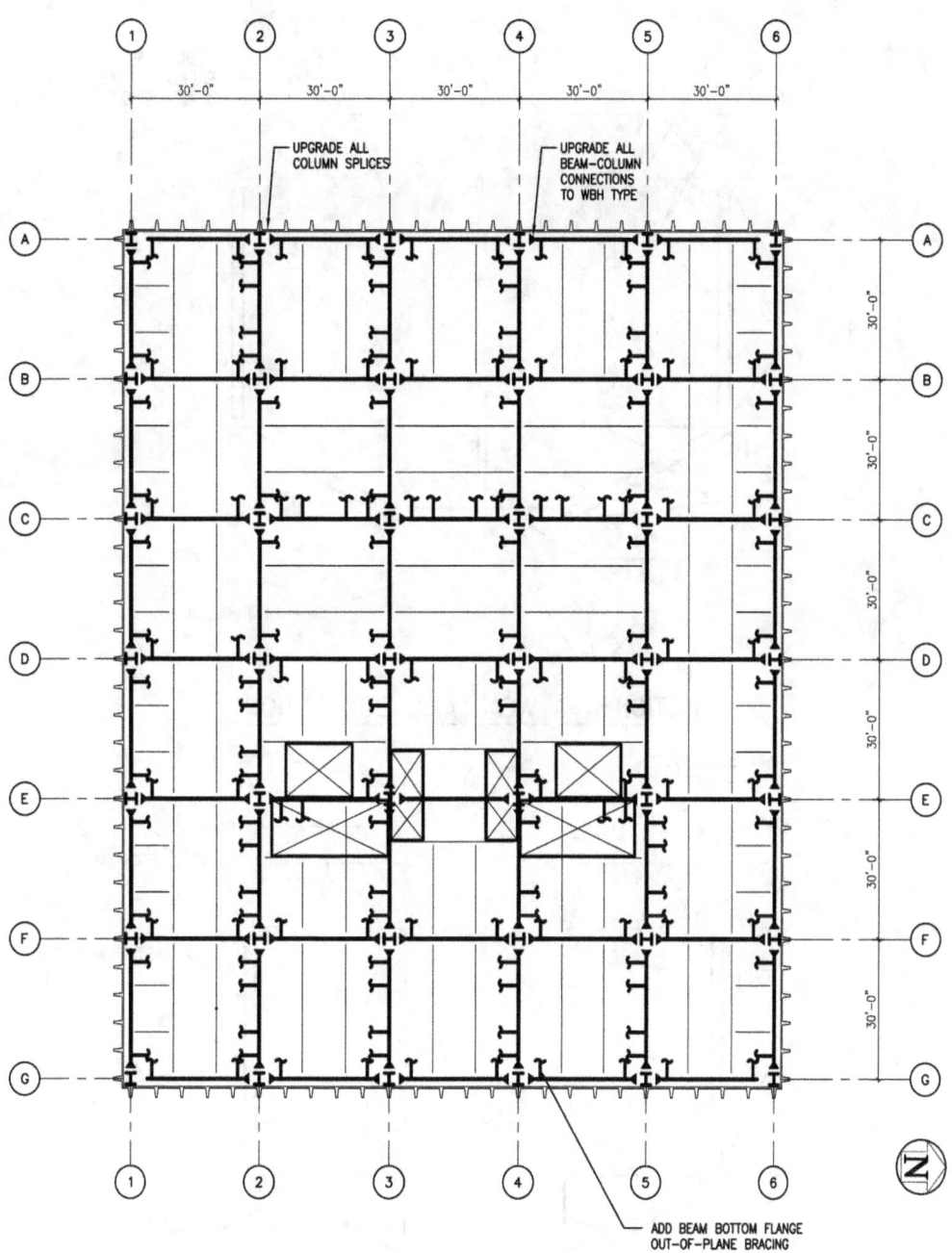

Figure 2-13. Connection Upgrade Plan

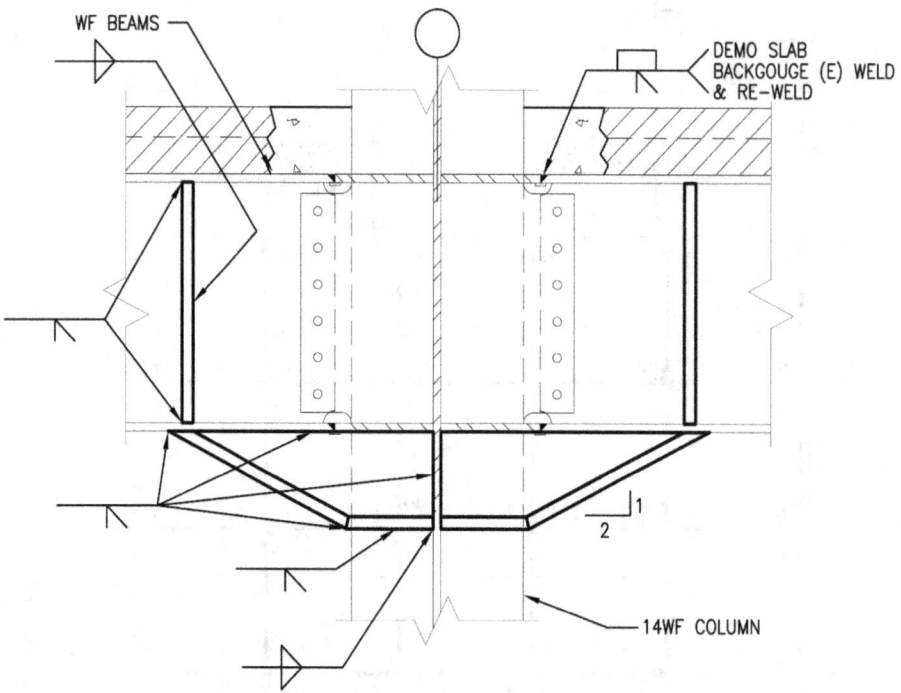

SECTION A-A WEAK AXIS CONNECTION

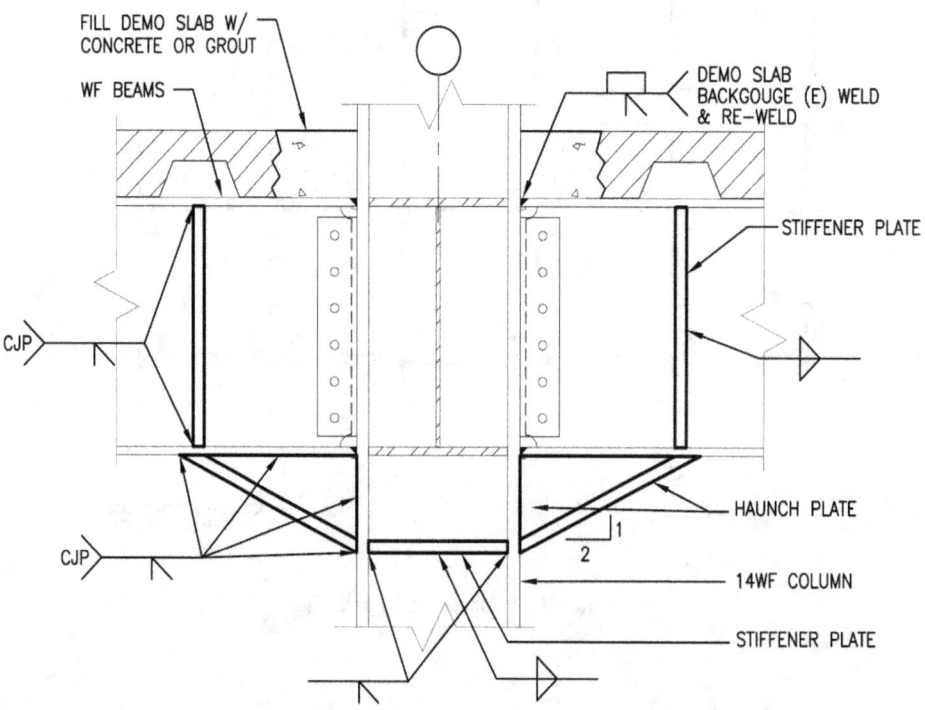

SECTION B-B STRONG AXIS CONNECTION

Figure 2-14. WBH Connection

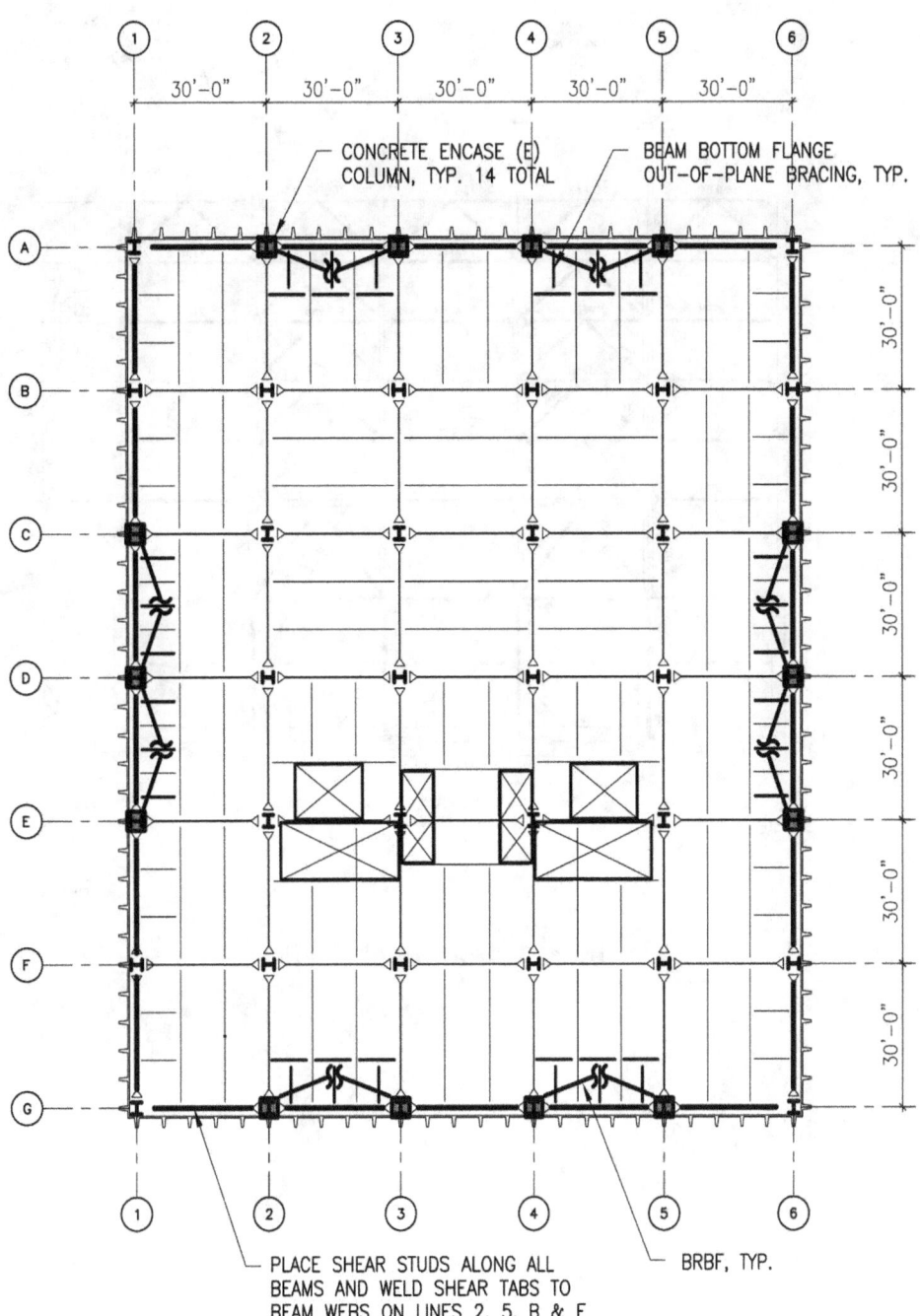

Figure 2-15. Exterior BRBF Plan

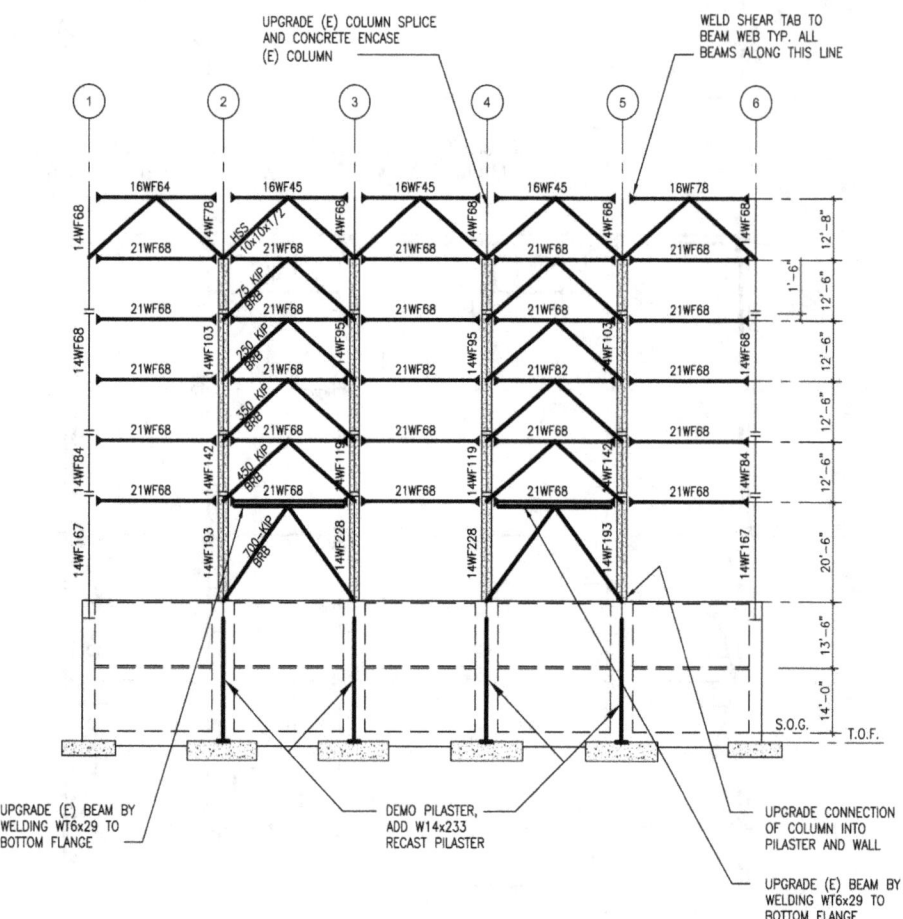

Figure 2-16. Exterior BRBF Column Line G

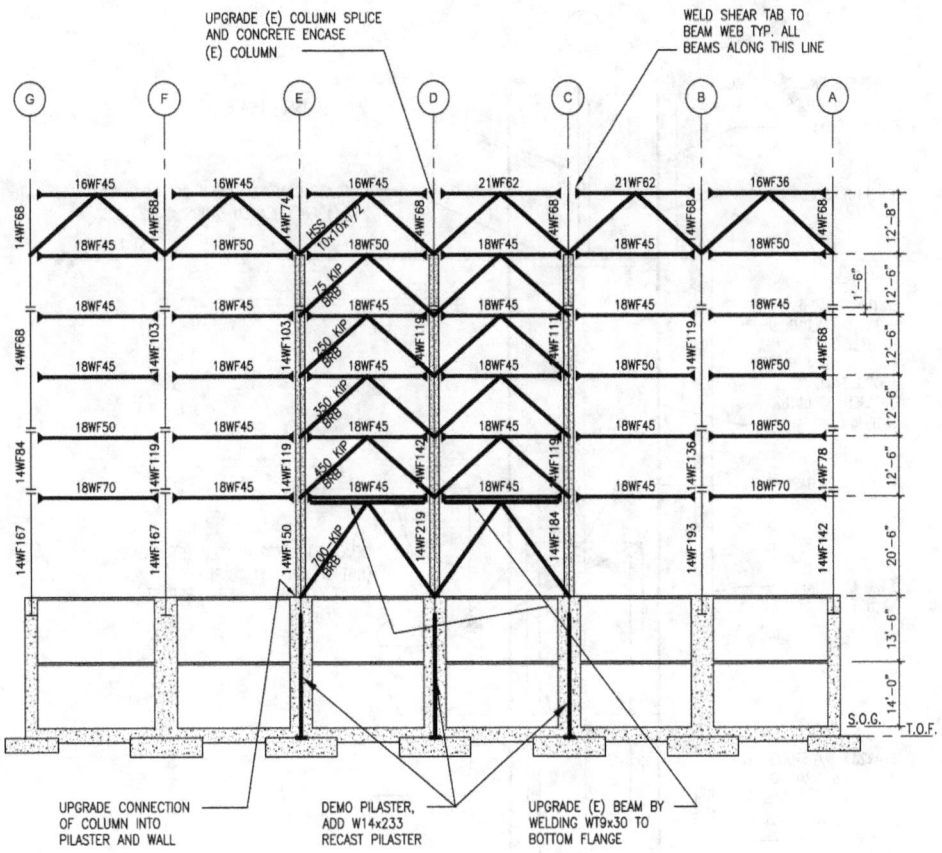

Figure 2-17. Exterior BRBF Column Line 1

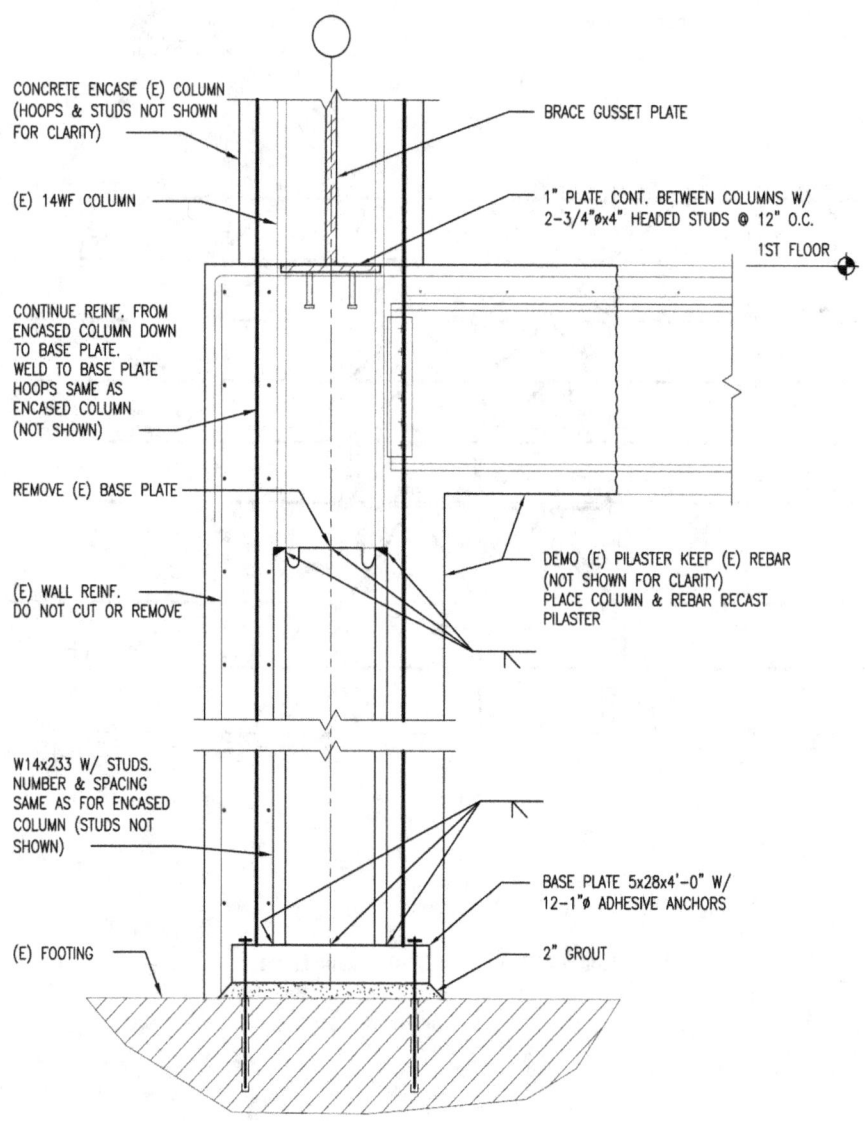

Figure 2-18. Connection of Existing Column into Basement Wall

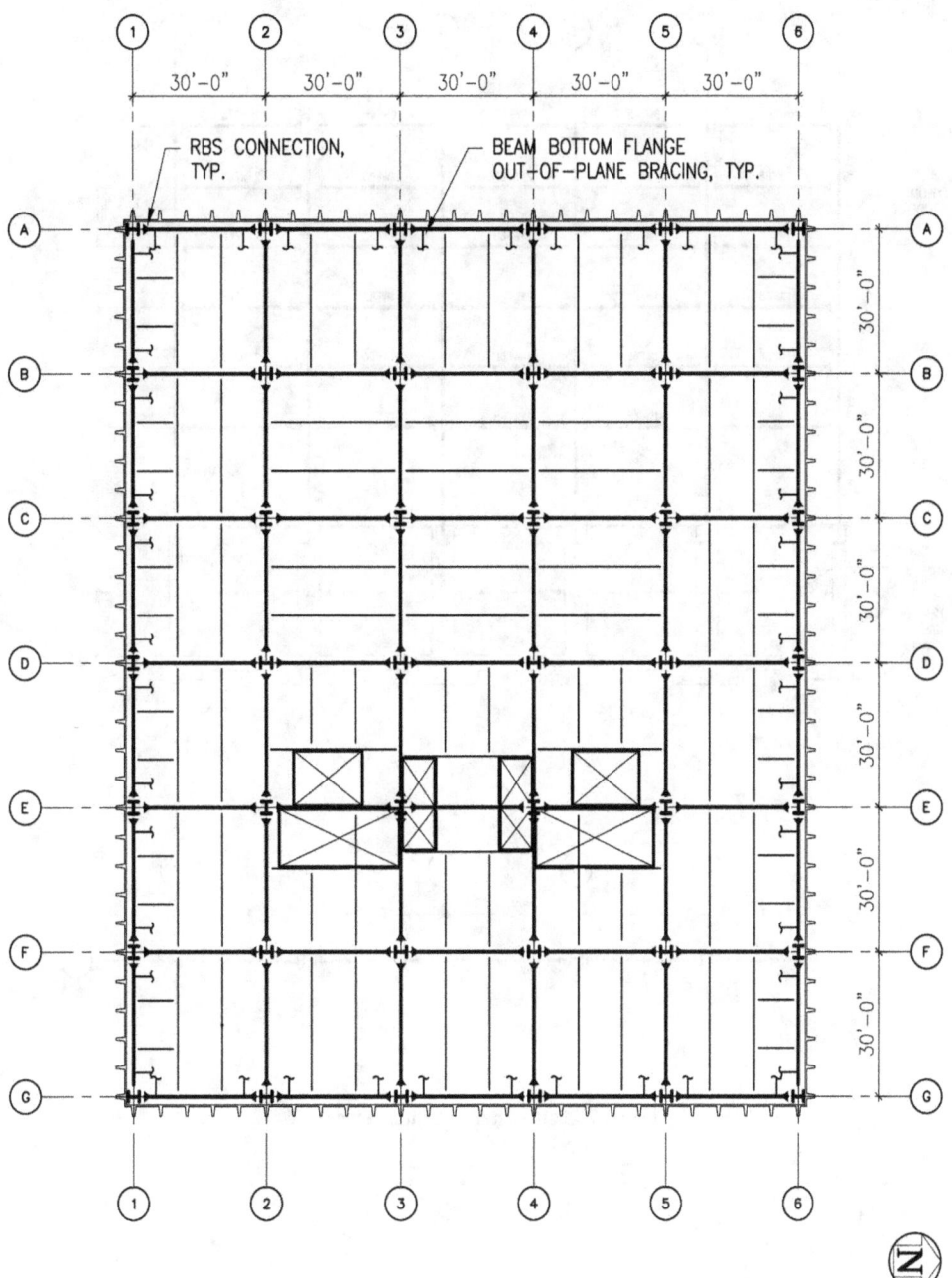

Figure 2-19. Plan of the Seismically Re-detailed Frame

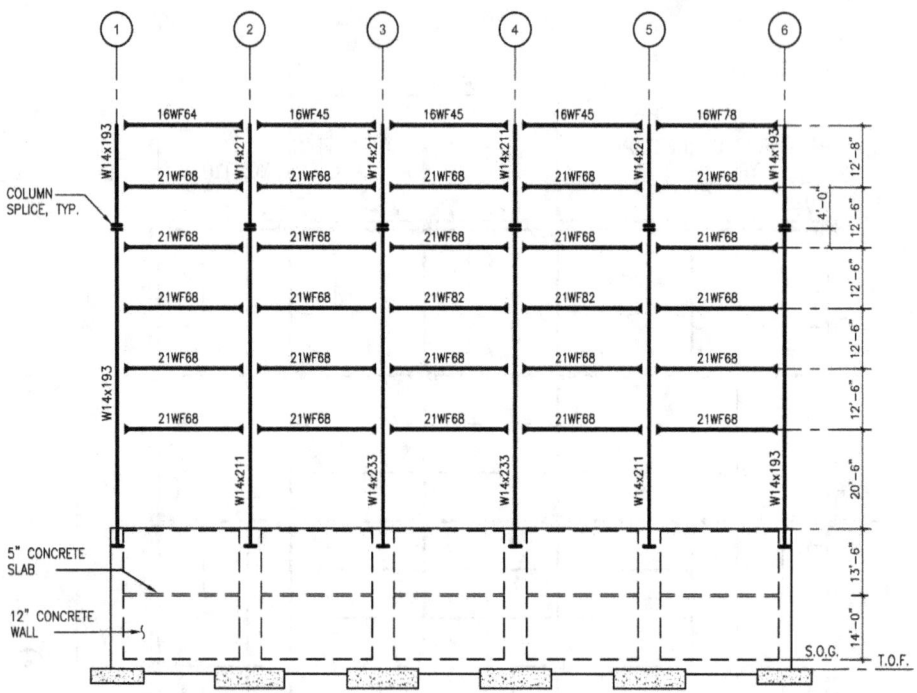

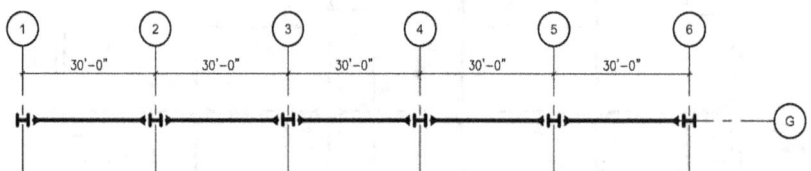

Figure 2-20. Re-detailed Column Line G

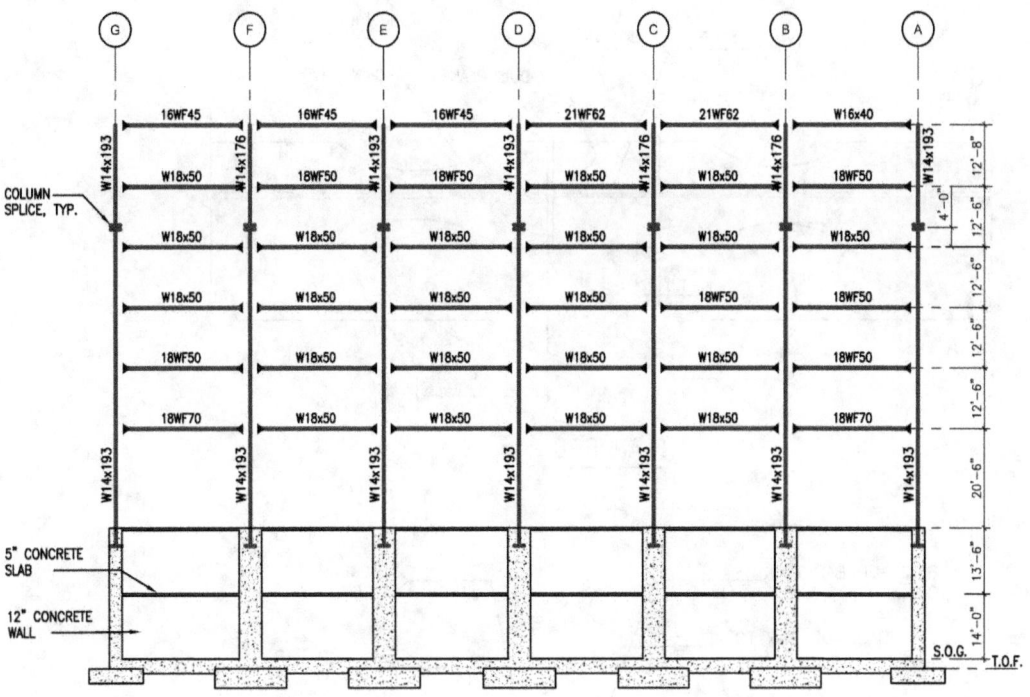

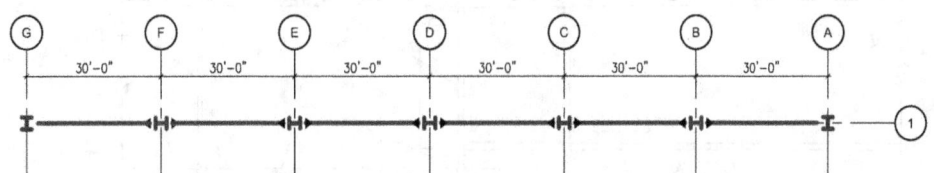

Figure 2-21. Re-detailed Column Line 1

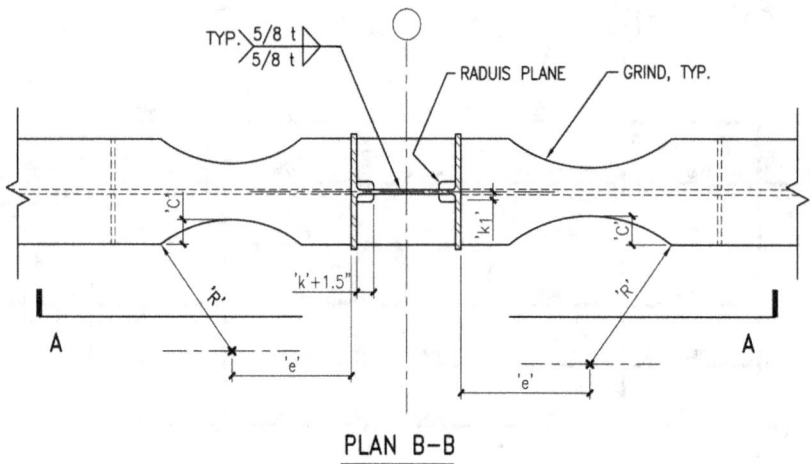

PLAN B-B

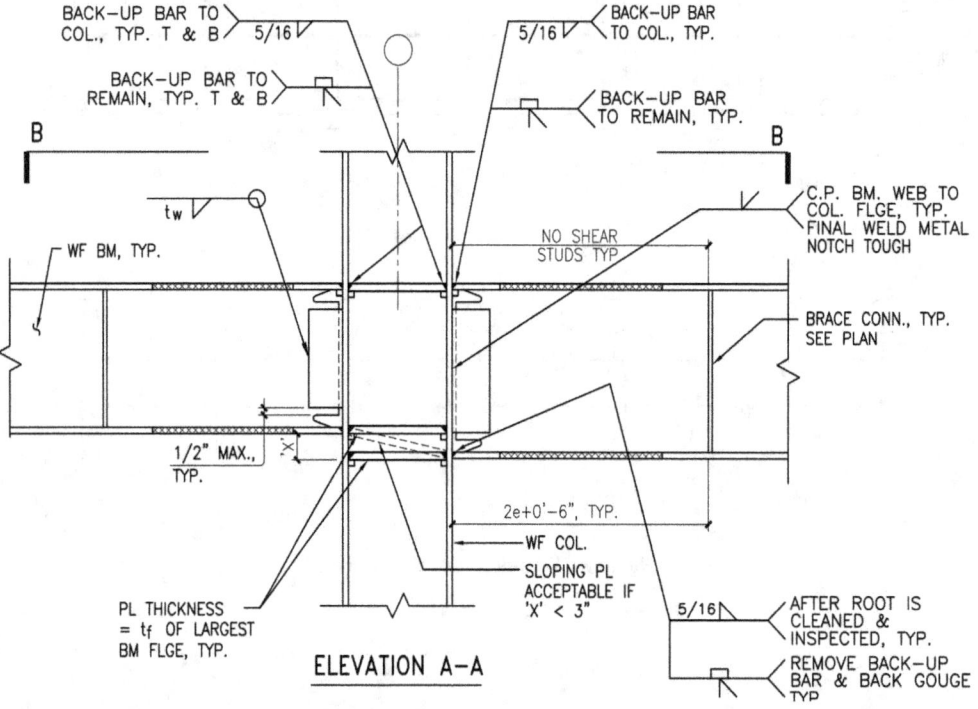

ELEVATION A-A

Figure 22. Reduced Beam Section (RBS)

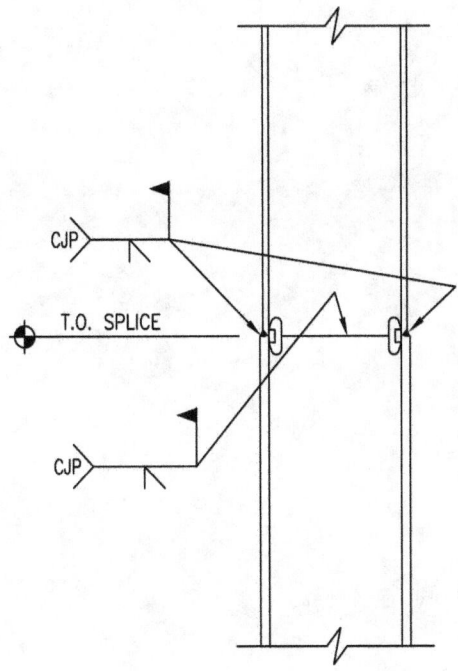

Figure 2-23. Column Splice Re-detailed

3   Elements That Were Strengthened to Improve Earthquake Resistance

3.1   Introduction

This chapter presents analytical results and conclusions regarding the blast responses of critical structural elements in the steel building. For more details regarding the actual calculations performed, the reader may refer to Appendix D. The discussion and the procedures used herein are not intended to serve as a model example of blast analysis or as a guide to blast design/analysis.

The explosive size and location were selected to be consistent with the Murrah Building truck bomb, as described in the FEMA 277 report (Reference 1). The explosive location was referenced to the cladded face of Column G3 of the steel building exactly as the explosive was referenced to the face of Column G20 of the Murrah Building. Chapter 4 discusses the system-wide progressive collapse scenarios that may ensue following the direct blast damage.

The level of the analytical effort for this study was intended to be consistent with that reported in FEMA 439A (Reference 2), focusing on using personal computer (PC) single degree of freedom (SDOF) models of the individual structural components to assess the blast response. The SDOF analysis and results are described in Sections 3.2 through 3.7. At the study's outset, it was clear that the extent of field and experimental experience with blast response of steel elements was substantially less than that for concrete elements, against which most of the SDOF models have been calibrated. This made SDOF modeling less reliable than for the FEMA 439A study. The SDOF computer models used in the blast response analyses and general limitations of those models are summarized in Section 3.2. The FEMA 439A and Appendix D contain a more extensive discussion of these models. Following the general model descriptions are the discussions of analyses for the original building, each upgrade scheme, and the re-detailed scheme described in Chapter 2. The response of the original structure serves as a baseline for comparison with the various upgrades.

Coincident with this study, a combined series of field explosive tests and analytical modeling of steel elements was undertaken jointly by the Defense Threat Reduction Agency (DTRA) and General Services Administration (GSA). Some of the analytical modeling experience gained from that program was brought into this project. Several nonlinear finite element analyses were performed for comparison with the SDOF modeling. These analyses are described briefly in Section 3.8.

For use as further validation of the analytical modeling results, the American Institute of Steel Construction (AISC) sponsored a field explosive test of a column similar to Column G3 in this study's building. That test column was also tested for its residual axial load capacity in its post-blast condition. A brief summary of those test results is provided in Section 3.9.

3.2 SDOF Computer Model Summaries

The PC-based engineering level analytical models used for the blast analyses are the U.S. Government-owned computer programs *ConWep* (Reference 30), *Span32* (Reference 32), *BlastX*, and *WAC* (Reference 34). All, except *BlastX*, were also used in the FEMA 439A study (Reference 2) and are described in some detail in that report and to a lesser extent in Appendix D. The programs were developed and are maintained by the U.S. Army Engineer Research and Development Center (ERDC) Geotechnical and Structures Laboratory (GSL) and the U.S. Army Corps of Engineers (USACE) Protective Design Center (PDC). They are restricted to use by U.S. government agencies and their contractors. Each code undergoes periodic revision and enhancement.

These programs were originally developed to support the design of reinforced concrete protective structures. The programs are therefore typically intended to provide life-safe protective designs and may be considered to be "design-conservative." The programs may overestimate the response of structural components, which for protective design basis is likely to be acceptable. However, this posed some challenges in attempting to create a baseline assessment of the original structure's actual performance. Judgment was exercised by the project team members when interpreting the computer output to determine which, if any, members actually failed under direct blast effects.

As discussed further throughout this chapter, the analytical models used in this study are limited in their ability to predict the response of structural components very close-in to an explosion. Generally, very close-in refers to scale ranges less than 1.0 ft/lbs$^{1/3}$. The term *scaled range* refers to the distance from center of the explosive charge to the face of the structural component, divided by the cube root of the charge weight. For such close-in events, the blast loading is very intense. Within that range, the loading gradient will vary significantly along a structural component. Additionally, the response of structural components at the very close-in scaled range is typically dominated by material response rather than structural element response. Therefore, simple structural mechanics assuming flexural behavior will not capture the response in the very close-in range.

Additional discussion concerning the computer programs and the limitations of their use for this study is presented in Appendix D.

3.3   Simplified Blast Analysis of Original Building

3.3.1   Explosive Threat

Modeled after the Murrah bombing, the explosive threat is an off-axis detonation of a 4,000-pound TNT-equivalent explosive near Column G3. Column G3 is a 14WF228 steel wide-flange section that is covered by decorative marble cladding. The center of the truck bomb is offset normally 12 feet 6 inches from the face of the cladding and offset laterally by 7 feet from the center of the column. The 12-foot 6-inch normal offset from the face of the column cladding corresponds to the offset from the face of the Murrah Building concrete column. The general scenario is illustrated in Figures 1-1 and 3-1.

3.3.2   General Modeling Considerations

Unlike the FEMA 439A study for which the actual response of the original, unimproved building to the attack was known, the baseline condition (the response of original building to the explosive threat) for the steel building was unknown and had to be determined analytically. Since the scope of this study was initially limited to simplified analysis techniques similar to the FEMA 277 and FEMA 439A studies, the first attempt to determine the baseline condition used SDOF models. As previously mentioned, experimental data on steel frame structure responses to blast effects are very limited. It was therefore not possible to establish definitively the level of confidence in the simplified analyses that was present in FEMA 439A. Therefore, the nonlinear FEA described in Section 3.8 was used as limited validation of the simplified analyses. Those results were then validated using the blast test summarized in Section 3.9.

Much thought was given to defining the loading areas to be used in determining the *ConWep* loading functions for individual column and beam components. It was decided to apply loading only to actual member dimensions, instead of including some assumed wall or cladding surface widths that would attract more total load. This assumption was based on the judgment that architectural cladding integrity would not be maintained long enough during the blast event to permit load transfer from the cladding to the steel elements. However, the presence of the cladding for airblast clearing effects was considered in defining the impulse applied to the columns. Equivalent uniform loading values for the frame elements that were calculated in *ConWep* were used in the *Span32* analyses.

3.3.3 Columns

The first story G-line columns are clad with marble. For the simplified analyses, blast loadings corresponding to the 12.5 foot normal and 7.0 foot lateral offset from the face of Column G3 cladding were applied directly to the steel columns. The first story G-line columns were completely exposed on all four sides because of the one bay building set-back at the first story. Consequently, those columns were susceptible to loading in both the strong and weak axis directions. The second-story and higher G-line columns were flanked by adjacent wall cladding. They were assumed to be loaded only in their strong axis directions (the direction perpendicular to the face of the cladding), with a slightly different standoff (normal to the wall) of 11.75 feet because of the geometry of the cladding.

For all columns in the SDOF analysis, the assumed support conditions were fixed-fixed. The loading function was based on an assumed uniformly distributed load over the length of the element and the resistance function was derived assuming a three-hinge flexural mechanism. This resistance function does not take into account localized failures, local buckling, or lateral-torsional response – all of which may occur particularly when the column is located close-in to the blast. Each element was also analyzed in isolation from all other elements in the building. The total response was then aggregated outside the models using appropriate judgment.

3.3.3.1 G-Line Columns

Due to its proximity to the bomb (scaled range of approximately 0.9 ft/lbs$^{1/3}$), the first-story Column G3 incurs the most intense loading. Figures 3-2a and 3-3b show the pressure and impulse distributions over the G3 column face. Although the peak pressure applied to the column is about 10,000 psi (Figure 3-2a), the equivalent uniform load for the entire column story height is represented with a peak pressure less than 6,000 psi (Figure 3-3). The simplified analysis technique with an equivalent uniform load most likely does not accurately capture the response for this column because of the actual highly non-uniform loading over its height. This can be seen in the results of the nonlinear FEA reported in Section 3.8 that show a maximum response occurring in the lower quarter of the height of the column, not at mid-height. It is therefore expected that the response of the first-story Column G3 may be considerably more than predicted by the simplified SDOF analysis technique. The simplified analysis technique is considerably more accurate for structural components that are further away from the bomb and thus have more uniform loading distributions – the remaining G-line and F-line columns.

In this study, the response of a structural component, such as a column or girder, is primarily quantified by support rotations (θ) and/or mid-length deflections (Δ). A support rotation is not to be confused with a joint rotation, which is a common term in structural analysis of a structural frame. A support rotation is the angle of rotation incurred at the end of a structural component as it interfaces with a support or joint. The rotation occurs due to the formation of a plastic hinge at the support. Figure 3-4 presents the maximum strong axis deflections and associated support rotations computed for the G-line columns, based on the SDOF analyses. Also, scaled range (z) values for the first story columns are presented. The tic marks on the girders of the frame in Figure 3-4 (and similar figures that follow) represent locations of the intermediate beams that provide some lateral support to the girders. Figure 3-5 presents the weak axis responses for the G-line first story columns. The weak axis loading will be minor because of shielding provided by the exterior wall panels for the second story and higher columns. Therefore only the first story columns are shown in Figure 3-5.

In an attempt to represent more accurately the response of the ground story columns, the first story G-line columns were analyzed for combined (biaxial) response, as opposed to decoupling the strong and weak axis responses as discussed in the previous paragraphs. For the biaxial analysis, the projected column width perpendicular to the line of sight from the center of the column to the center of the bomb was used to determine loading area. Cross-sectional properties about the axes corresponding to the rotated column cross-sections were used to compute resistance functions. Figure 3-6 presents maximum biaxial responses of the first-story G-line columns.

Figures 3-4 through 3-6 show that first story G-line column responses are dominated by weak axis response. Biaxial response is only slightly greater than the weak axis response considered in isolation. The responses are relatively low compared to the level of response that might be expected from such an event. The maximum deflection occurring at Column G3 for the biaxial response is estimated to be 1.8 inches, which is much less than the 12-inch displacement predicted by the FEA analysis and the 4-inch permanent displacement observed in the blast test. This is most likely due to the non-uniformity of actual loading pressures due to the column's proximity to the blast. The column has a much greater and more intense loading just above its base, which leads to localized deformations in that region. FEA and testing can capture that, while the SDOF analysis cannot capture that unless the resistance function input explicitly assumes that failure mechanism will occur and estimates how it will occur. Such a resistance function could not be developed without having a baseline of numerical data.

Loading on columns other than the first story of G3 is much more uniform and less intense. Responses of the other G-line columns that are computed with the simplified analyses can therefore be considered to be reasonable estimates, with a maximum computed deflection of just over 3 inches for Column G2. As an example, the peak equivalent uniform pressure (biaxial) applied to the next closest column (first-story Column G2) is approximately 2,400 psi, whereas the actual peak pressure distribution on Column G2 varies from approximately 1,900 psi to approximately 3,000 psi.

Due to the orientation of the columns with respect to the bomb location, Column G2 incurs a more direct weak axis blast loading than Column G3. Consequently, Column G2 experiences a greater response with respect to its weak axis. Similarly, Column G4 experiences a greater weak axis response than Column G3. The angle of incidence of a surface (such as a face of a column) greatly affects the magnitude of the reflected blast pressure and impulse. The web surfaces of Columns G2 and G4 approach normality with the line of sight to the bomb much more so than the web surface of Column G3. Consequently, Columns G2 and G4 undergo primarily weak axis response. However, reflected blast pressure is greatest on the flange face (rather than the web surface) for Column G3; thus, strong axis resistance is mobilized. Therefore, it is not surprising that the analytical procedure used in this study predicts greater overall response for Columns G2 and G4 than for Column G3. Again, the analytical procedure does not capture the effect of the intense blast loading in the lower portion of G3.

Due to the presence of the exterior wall, weak axis loading is not significant for the second story and above; only strong axis bending is of interest for the upper stories.

The simplified analysis results are substantiated by comparing maximum deflections of the third story Column G3 computed by FEA and *Span32*. As described in Section 3.8.3 and presented in Figure 3-32, the maximum response of the column in the FEA is approximately 0.25 inches. *Span32* computes a response of approximately 0.34 inches. Considering the many assumptions used in both the airblast and structural modeling, both analyses predict a response of approximately 0.3 inches.

Based on the SDOF analysis results alone, it is not evident that any G-line columns will experience failures (i.e., rupture or excessive deflection). However, a column must have adequate ductility to form the three-hinge mechanism assumed for the analysis and undergo plastic deformation. For the subject building, a column splice is located 18 inches above the second floor slab. In the seismic evaluation the columns splices were found to be deficient because the webs of the columns were not welded to each other and the flanges were only welded with partial joint penetration welds. Because of the lack of web attachment, the flange partial joint penetration welds must carry both the shear and flexure demands from one column to the other. To determine if this connection was adequate, the moment and shear demands in the splice were calculated by assuming that plastic moments occurred in the column at midspan and both ends. This is the assumed limit state used in the SDOF blast response analyses. It was determined that the demands in the column splice welds from that condition exceed the welds' capacities. Consequently, the splice will not permit the development of plastic moments in the column at the splice location. The demand on the splice weld connection to accommodate plastic hinge formation in the column is over 20 percent greater than the weld connection capacity. Therefore, it is likely that the column would fail at the splice before it could respond as predicted by the SDOF model. The failure may be brittle. Consequently, the gravity support capability of Column G3 is conservatively estimated to be lost just above the second floor slab at the splice, and the column will be deflected into the building.

3.3.3.2 F-Line Columns

Considering the F-line, only the first-story columns are exposed to direct blast effects and only Columns F1 and F6 are not bounded by adjacent wall panels. Unlike the G-line columns, F-line columns are oriented such that the weak axis of each column faces the building exterior (bomb side). Wall panels essentially shield Columns F2 through F5 from strong-axis loading. First-story Columns F1 and F6 are exposed to strong-axis loading, but their responses are dominated by weak axis loading. Figure 3-7 presents analysis results. The largest maximum response of the F-line columns is incurred by Column F3, which has a maximum response exceeding 1½ inches. Those deflections are small enough to conclude that those columns will not fail due to direct blast effects.

3.3.4 Girders

The girders on the west side of the building (G-line) are shielded by wall panels, as shown in Figure 3-8. Depending on the response and debris field generated by blast effects on the wall panels, the G-line girders may incur debris impact loading. Limited experiments and high performance computations have indicated that structural member response to cladding debris field impact loading is similar to direct airblast loading when the loading is relatively intense. However, at some distance from the bomb, the cladding will reduce loading effects on structural elements. For this study, the cladding was assumed to be destroyed, and girders were assumed to be directly loaded by airblast along their weak axes, which should generally over-predict girder response based upon the debris field impact load tests.

The girders span approximately 30 feet between columns, but are laterally braced by perpendicular beams at a spacing of approximately 10 feet. The top flanges of the girders are connected to the concrete filled metal deck with puddle welds between the metal deck and beam flanges, which would provide some restraint against the beams' top flanges moving inward, though not nearly the amount of restraint that would be provided if welded shear studs were present. That restraint would cause the beams to deform in a lateral-torsional manner, instead of a pure flexural manner. However, this effect could only be postulated and could not be quantified because of the absence of any test information or detailed computational studies. It was therefore chosen to conservatively assume that the girders' top flanges were not restrained. Therefore, girder clear span for response calculations was assumed to be 10 feet and three-hinge mechanisms were assumed to occur in the simplified SDOF analysis.

Figure 3-9 presents results of the *Span32* analyses. The values shown for each column-to-column span in Figure 3-9 represent the maximum response of the one of the three 10-foot girder lengths within that 30-foot span that has the greatest response. That 10-foot length is typically the girder section closest to the bomb. The "X" on each girder in Figure 3-9 indicates the 10-foot length corresponding to the presented response values. As with the column analyses, the individual components (10-ft girder sections) are analyzed as independent members with stationary, fixed supports. In reality, there is a global response such that the girders are displaced as the columns and perpendicular beams deform.

As shown in Figure 3-9, the greatest response in the simplified analysis is over 60 inches of displacement horizontally into the building, computed for the second floor 10-foot girder section that is closest to the bomb, between columns G2 and G3. The next largest response is over 30 inches for the girder between columns G3 and G4. The responses of all other girders are computed to be less than 11 inches. Because of the large rotations associated with these displacements, weld failures in the directly welded flange connection will result in failure of the second and third floor girders' connections to Columns G2, G3, and G4. A response limit of 12 degrees for steel beams is generally accepted by DOD blast-related design manuals, but does not directly account for connection details susceptible to fracture. The blast effects community accepts the 12-degree support rotation as representing heavy damage in a severe event, but simply assumes that the connection will survive to allow the rotation to occur without collapse. Currently, little is known regarding the behavior of the steel beam-column joint under very large deformations.

These responses appear to be excessive when compared to the columns that are closer to the blast. However, it must be noted that the girders are being loaded primarily along their weak axis, which has substantially less stiffness and strength than the weak axis of the columns, while attracting more blast pressure due to the greater surface area than the columns. It is noted once again that these postulations represent the conservative case in which the restraining effects of the floor slabs is neglected.

Only the second floor girders between Columns G2 and G4 experience support rotations greater than 12 degrees. In reality, the top flange of the girder is partially restrained by the puddle weld connection to the corrugated steel decking of the floor slab. As stated before, this partial restraint is not considered in the reported response deflections, and would cause some reduction in response.

Given the uncertainties with respect to connection integrity, it was concluded that the second floor girders between G2 and G4 would definitely be lost and the third floor girders between G2 and G4 may also be lost. There is again uncertainty in these models due to the lack of any field tests for girders loaded in their weak axes and with welded moment connections to the columns.

3.3.5 Floor Slabs

The floor slab system for the building was described in Section 1.4.3. As previously mentioned, the first (ground) floor and basement floor slabs are cast-in-place reinforced concrete and have some degree of structural integrity with supporting girders and walls. For the upper levels of the building, floor slabs are comprised of concrete-filled metal deck diaphragms. Because of known characteristics of such diaphragms and because of the details shown on the building's structural drawings, it was considered to be unlikely that the welds connecting the diaphragms to the supporting steel girders would provide significant resistance to net uplift forces that would occur in the regions that were close to the bomb.

Again using the simplified modeling approach, considerable effort was devoted to determining the response of the floor slabs to the external detonation. The complexity of the airblast and debris loadings, as well as the diaphragm support conditions, makes modeling this response very difficult for the simplified design-oriented tools used in this study and even the sophisticated analytical tools such as nonlinear FEA. The computer program *BlastX* is a PC-based code primarily developed for modeling detonations that occur internally in buildings. *BlastX* is capable of considering the contributions of shock reflections from surrounding surfaces. A brief description of *BlastX* is provided in Appendix D.

If the blast wave was allowed to propagate into the building, there would be reflections in addition to those assumed if *ConWep* was used. To estimate the extent of damage to interior floor slabs, a *BlastX* model was developed to estimate interior pressures in the front bays of the building at all floor levels. For this evaluation, it was assumed that window glazing and most of the exterior walls did not exist and thus did not attenuate blast propagation into the rooms. Based on observations in other field testing (Reference 42), this is a very conservative assumption. A net uplift pressure time-history was determined for each floor slab in the front bays of the building by subtracting the downward loading (blast plus gravity load) on the top of each floor for each bay from the upward loading on the bottom surface of each floor. The Wall Analysis Code (*WAC*) is an SDOF code much like *Span32,* with good capabilities for inserting user-defined resistance functions, and it was used to predict the response of the floor slabs. This general approach was also used in the FEMA 277 study of the Murrah Building. The total lost floor slab area in the Murrah Building was attributed to a combination of direct response to airblast and to progressive collapse effects. The conclusions concerning areas of slab loss due to direct blast loading could not actually be verified.

Applying the procedure described above, all top story floor slabs and the second through fifth story floor slabs between column lines 5 and 6 incur upward deflections of a few inches. The WAC analyses indicate that all other floor slabs were overloaded, experiencing very large deflections (slab responses typically did not achieve equilibrium during the analyses). The procedure produces large loads on the floor slabs, the lowest being on the floor of the farthest bay from the bomb in the building corner of the upper story. That floor is given a uniform uplift load corresponding to a peak pressure exceeding 12 psi and an impulse exceeding 45 psi-ms.

The large displacements and slab failures result from the blast infilling the building because of the assumed absence of exterior windows and walls. In contrast to the initial conservative assumption that glazing and walls do not attenuate the incoming air blast, it is well known from experimental research (Reference 42) that window glazing and light walls can significantly attenuate blast propagation. Therefore, it is likely that the initial assumption is very conservative, leading to an over-prediction of slab damage. The interaction of the blast loading with the response of the exterior cladding and windows, as well as the subsequent infill of any blast pressures is an extremely complex event that cannot be modeled accurately with such simple tools as those used here. Additionally, impact loading from the exterior wall debris can have a significant effect on the floor response. Consequently, this study cannot draw strong conclusions regarding the extent of direct blast-induced floor slab damage, but instead focuses on the frame response.

It is reasonable to assume that at least the floor slabs of the front bays between Lines 1 and 5 on the second floor near the location of the vehicle bomb will be destroyed by direct blast and debris impact effects. Because the building's façade is set back at the first story, those slabs do not have cladding to shield the upward blast pressure on them. Also, the top surfaces of the slabs are shielded by the exterior walls of the second story. Thus, the net upward forces on the second floor slabs would be large. Therefore, for those slabs the simplified analysis was judged to be acceptable.

3.4   *Simplified Blast Analysis of Moment Frame Connection Upgrade Scheme*

The moment frame Connection Upgrade Scheme provides improved beam-column connections and column splices (see Section 2.2.3). However, the enhancement of the beam-column connections is difficult to quantify, and the simplified analysis techniques used in this study assume that individual structural components already have significant support fixity. Therefore, it was conservatively assumed that the girder response would be similar to the original building, with the second floor girders being lost between Lines 2 and 4, and the third floor girders being damaged but not lost.

The improved column splice enhances the column's blast resistance. Unlike the baseline condition discussed in Section 3.3.3.1, the upgraded column splice will allow the 3-hinge mechanism to form in the column. Thus, the predicted maximum deflections reported in Section 3.3.3.1 for the baseline structure's columns are not compromised by column splice failure in this connection upgrade scheme. Those deflections are considered to be representative of the responses in this upgraded scheme and it could be inferred that the column would not be lost to direct blast effects.

3.5 Simplified Blast Analysis of Exterior BRBF Upgrade Scheme

3.5.1 Encased Columns

The Exterior BRBF (Buckling-Restrained Braced Frame) upgrade scheme primarily consists of providing the Buckling Restrained Braces (BRB) between several bays up the height of the building and encasing the columns the braces attach to in concrete (see Section 2.2.4). Column encasement increases the loading surface area, so that the total blast load applied to each column is greater than that of the original building design. The encasement increases column flexural resistance, because of increased section properties. More importantly, the increased mass from the concrete increases the inertial resistance, which contributes greatly to increased blast resistance.

Each encased column was analyzed as a reinforced concrete member with steel reinforcement comprised of the steel section and all added reinforcing bars. Figure 3-10 presents the results of *Span32* analyses of strong axis response for the first and second story encased G-line columns. In comparison to the response of the original building, maximum deflections are lower, not exceeding approximately 0.6 inches. Because of the low level of response in the lower stories, it was not deemed necessary to analyze responses of the third and higher stories. Similarly, Figure 3-11 indicates that weak axis response of each first-story G-line column is very low, with maximum deflections of less than 0.5 inches.

The 0.6-inch deflection mentioned above is for encased Column G3. The loading on this column is very intense and non-uniform. *ConWep*, which incorporates currently-accepted empirical breaching (a breaching failure means that the concrete in a localized region would be destroyed before the member itself would have a chance to respond to the blast) calculations for typical <u>reinforced concrete</u> elements (see FEMA 439A), indicates that Column G3 would have to be approximately 60 inches thick to prevent severe damage to the reinforced concrete. The encased columns are only 26 inches thick. The *ConWep* calculations would imply that this would occur at the first story G3 Column.

However, unlike the reinforced concrete members tested to establish the breaching curves in *ConWep*, the encased column has the large steel wide-flange section embedded within it. There is a lack of experimental data for blast response of steel columns, but the lack of experimental data for response of close-in blast on concrete-encased steel columns is even greater. The breaching equations do not include the effects of reinforcing steel, particularly the substantial amount of confinement that is required by code and provided by the transverse reinforcement, since experimental data generally indicate little effect on the breaching resistance of reinforced concrete subjected to intense blast loading. Severe breakup of the front face (front of the encased steel column) concrete and spalling of the back face concrete is likely. It is reasonable to conclude that concrete-encased Column G3 will deflect less than the 11 to 12 inches computed by the nonlinear FEA for the original Column G3, but more than the 0.6 inches computed by *Span32*. It will likely survive well enough to maintain the required gravity load-carrying capacity to avoid collapse. Computed deflections presented for the other columns are again considered reasonable estimates of response.

3.5.2 BRBF Components

The responses of the diagonal braces in the BRBF to direct blast loading were also considered. Since the braces consist of structural steel sections embedded in concrete-filled tube sections, resistance functions for the BRBF members were developed and used in the "user defined" option of *Span32*. The tube and concrete section of each BRBF member terminates near its connection to the column. Thus, it is reasonable to assume that hinges will first form at the supports and then at midspan. The direct blast loading was estimated from *ConWep*.

From the *Span32* analysis, it was found that the blast did not damage the braces to the point where they would be lost. The maximum midspan displacement of the brace closest to the blast at the first story extending from column G3 was approximately 5 inches.

3.5.3 Girders

Because the connections between the beams and columns were embedded in the concrete encasement, there would still be some support for gravity loads due to bearing of the bottom flange on the 6 inches of concrete encasement on each side of the column. It was therefore assumed that while the beam-column connections of the third floor girders may again be damaged like in the other schemes, they would still remain. The second floor girders would be lost between G2 and G4. With the loss of the beam between G2 and G3, the braces would also probably be lost.

3.6   Simplified Blast Analysis of Interior BRBF Upgrade Scheme

3.6.1   Encased Columns

Similar to the Exterior BRBF upgrade scheme, the Interior BRBF upgrade scheme primarily consists of providing BRBF bracing and encasing interior columns in concrete (see Section 2.2.2). The same approach that was used for the G-line columns of the Exterior BRBF scheme (i.e., using the embedded steel column as reinforcement for the reinforced concrete columns) was used for the F-line encased columns of the Interior BRBF scheme. Figure 3-12 presents the results of the *Span32* analyses for the F-line first story encased columns. The computed deflections are considerably less than those computed for the original F-line columns. For example, the maximum deflection of the encased first story Column F3 is less than 25 percent of the original first story F3.

In this scheme, however, the G-line (and other exterior lines) columns were not strengthened, so that the vulnerabilities of the building exterior would be largely the same as those found in the analyses of the original building. Therefore, the post-blast condition of this scheme would be the same as for the baseline structure.

3.6.2   BRBF Components

Similar to the Exterior BRBF upgrade scheme, first story diagonal braces were assumed to be loaded by direct airblast over each member's dimensions. As with the F-line columns, upper story BRBF bracing is assumed to be adequately shielded by the exterior bay of the building.

3.7   Simplified Blast Analysis of Re-detailed Original Building Frame

The re-detailed frame primarily consists of resized columns, accompanied by Reduced Beam Section (RBS – "dogbone") beam-column connections (see Section 2.3). Additionally, the columns are rotated 90 degrees from that of the original building. The first story columns that are closest to the bomb will be dominated by weak axis response, as opposed to the strong axis response experienced for the original building. Strong axis loading is of little interest for all upper story columns, due to the shielding provided by the exterior wall panels. Figure 3-13 presents maximum weak axis responses of the first and second story columns. Figure 3-14 presents maximum strong axis responses for the first story columns.

Since the re-detailed columns have adequate splices, a premature splice failure in the column responses is not expected. Although the first-story re-detailed columns consist of larger W-sections than the columns of the original building, their rotated orientation results in a greater blast response at some locations for the given bomb size and location. However, responses are relatively small, remaining below 1.5 inches (except for over 2.3 inches at Column G3). The second story re-detailed columns consist of considerably larger W-sections than the original second story columns. Weak-axis bending resistance of the large re-detailed columns is greater than strong-axis bending resistance of the original columns; thus, the re-detailed columns will have less or approximately the same response as the original second story columns.

Another difference between the original frame and the re-detailed frame is that material yield strength of columns in a modern building designed for seismic resistance would also be greater (50 ksi nominal yield) than that of the original building (36 ksi nominal yield) of this study, leading to higher yield loads and thus greater resistance functions. The higher yield strength was considered in the SDOF analyses of the re-detailed structure.

The response of the second story re-detailed Column G1 is reported to be essentially equal to that of the original column.

The RBS connections affect the resistance of the G-line girders. The reduced sections do not affect the middle 10-feet length of each girder, but do affect the ends of the 10-feet lengths that attach to the column, because the weak axis moment capacity is significantly reduced at the location of the cut-away section. The reduced capacities affect the resistance functions used in the SDOF analyses in which hinges are considered to first form at the supports and then at mid-span, lowering the demand at which the first hinge forms. It was assumed that plastic hinges at the supports would have sufficient rotational capacity to allow a hinge to form at mid-span. As is the case for the original building analyses, the girders are taken to be loaded directly by airblast in the weak axis direction. Figure 3-15 presents the results of the analyses, and reports the deflection of the 10-feet length of girder that incurs the greatest response for each 30-foot bay. In many cases, the computed response of the girders is significantly greater (generally 2 to 4 times) than the computed values for the girders in the original building. Using a response-to-failure limit of 12 degrees of rotation, the second and third floor girders between Columns G2 and G4, as well as the second floor girders between Columns G1 and G2 will be lost due to direct blast loading.

3.8 *Nonlinear FE Analysis*

As mentioned in Section 3.1, a combined series of field explosive tests and analytical modeling of steel elements was undertaken jointly by the Defense Threat Reduction

Agency (DTRA) and General Services Administration (GSA) at approximately the same time that the SDOF modeling was being conducted. The analytic capability developed and validated for this work was used to support to this study by performing several non-linear finite element analyses (FEA) that could be used to assess the applicability of the SDOF analysis results.

Nonlinear FEA is often used to analyze structural components and systems subjected to blast effects when neither test data nor validated simplified engineering models are available. Specialists within the defense community have contributed significantly to recent advances in FEA, especially in high-fidelity physics-based (HFPB) methodologies.

Computer codes exhibiting HFPB capabilities are based on first principle physics rather than simplified engineering assumptions. Codes like *DYNA3D* (Reference 18), *FLEX* (Reference 19), *LS-DYNA* (Reference 20), and *PRONTO* (Reference 21) represent a more physics-based approach to employing finite element (FE) models than is commonly used for structural response calculations. For example, these FE models use material models that simulate the actual performance of concrete (e.g., rate effects, effects of confinement, softening, fracture, etc.). Other important attributes of these codes include contact algorithms, large-deformation geometry, and the ability to compute responses for highly damaged structures.

Computational results from HFPB FE calculations have been shown to compare well with test data. In addition, the calculations are often used in the defense community to create virtual data as alternatives to executing expensive field tests, such as mentioned in References 22-26. In the absence of test data and validated engineering models, HFPB FE methods can provide a good (and often the only) alternative to obtaining the response to blast load effects on steel structures. While this approach is very attractive, specialized training and expertise are required to generate accurate results, and erroneous results are sometimes difficult to detect, which requires extensive checking of results.

An analysis using an HFPB model developed using LS-DYNA was performed on the original steel baseline structure (i.e., no strengthening measures included) to provide an estimate of the anticipated structural response to a terrorist attack similar to the Murrah Building attack, as described in Section 3.3.1. The details of the analysis are included in Appendix D. In summary, this FEA of the frame indicates that it is not likely to collapse as a result of blast effects represented by this simulation. Although the primary column was significantly damaged, the structural system appears robust enough to resist a progressive collapse type of failure.

3.9 Field Testing Conducted to Validate Column G3 Response

While this study was underway, the American Institute of Steel Construction (AISC) sponsored a full-scale blast test of a column similar to Column G3 of the original building. Following the blast test, that column was tested to determine its residual axial load capacity. The tests and their results are summarized below. A full narrative on the column tests can be found in References 35 and 36, and are reprinted in Appendices F and G, respectively.

The column tested was a W14x233 section of ASTM A992 steel, with a clear height of 18 feet 9 inches (similar to the clear height of Column G3). The W14x233 differs only slightly from the 14WF228 Column G3. Typical cross-section parameters are only 2% larger. The major difference between the test column and original building's Column G3 is steel strength. Column G3 in the original building was made of A36 steel, which for a Group 3 section has an expected yield strength of 44 ksi, according to FEMA 351 (Reference 7). The expected yield strength for A992 steel is 55 ksi, so the test column's yield strength is 25% greater than the original building column.

The column was set in the test reaction structure such that rotation and displacement would be restrained at both ends, to simulate the "fixed-fixed" support conditions assumed in the SDOF analysis. To simulate the cladding debris loading, a 4-inch brick encasement was placed around the full height of the column. The encasement mass is similar to the 2-inch marble cladding around the building's first story columns, providing for a similar impact load.

Due to constraints with the test setup, the initial gravity load could not be applied. This should not be a significant concern. Because the blast load and the column's response to it occur so rapidly, the column will not respond to the gravity load applied to it until well after the blast response has occurred. Additionally, the estimated axial stress on the column is only 11% of its expected yield strength, so its flexural capacity should not be significantly affected by it. The nonlinear FEA illustrated this. Therefore, it is not unreasonable to test the blast response and then separately test to determine the axial load capacity.

The explosive charge used in the test was 4,860 pounds of ammonium nitrate-fuel oil (ANFO), which approximates the 4,000 pounds of TNT-equivalent used in the Murrah Building bombing. The charge was placed at the same location with respect to the test column and the same height above the ground as used in this study. Figure 3-16 shows a picture of the test column and explosive charge.

Figures 3-17 and 3-18 show the post-blast condition of the column. The blast completely destroyed the cladding. The column sustained a permanent strong-axis centerline displacement of 3.75 inches at 42 inches above the base. The flanges of the column folded in from the web, with permanent displacement of the flange tip further from the bomb being 6.38 inches and the flange tip closer to the bomb being 4.50 inches, both occurring at 21 inches above the base, leading to net displacement of 2.63 inches and 0.75 inches respectively. The column also sustained a permanent weak-axis displacement of 1 inch. As can be seen, the damage is minimal, especially given the large charge weight. From this it appears that the column would still have substantial gravity load carrying ability following the blast.

Following the blast test, the column was tested to determine its axial load capacity in the post-blast deformed condition. The column was prepped and placed into a load frame, with essentially pinned-pinned end conditions. This is a conservative boundary condition due to the limitations of the test setup. The column axial capacity was found to be approximately 1,700 kips, 62% of the computed axial capacity for an A992 W14x233 with an effective length of 19 feet. The observed failure mode was buckling about the weak axis, which caused the column to hit the side of the load frame.

The axial load on the column in the building after the blast was estimated to be approximately 330 kips (considering full dead load and 25% of the design live load). Assuming that the capacity of the post-blast column is 62% of the calculated capacity of an undamaged column, the capacity of the blast-damaged original first-story G3 column (of A36 steel) would be 1,400 kips. Based on that, it is reasonable to assume that, while column G3 would sustain significant damage, it would not lose its ability to support the gravity loads from the floors above.

3.10 Discussion of Analytical Modeling

This study has highlighted some of the limitations in present analytical modeling techniques for the responses of steel-framed structures to blast loading, particularly in design-oriented SDOF techniques. Those techniques have been validated for reinforced concrete structures by comparison of numerical results to observations made in many years of field explosive testing of concrete structures that simulated defensive bunkers comprised of significant slabs of heavily reinforced concrete. The SDOF models have become quite accurate for analysis of reinforced concrete component response that is dominated either by flexural action along traditional yield lines, localized shear effects at supports, or *brisance* (or breaching – complete shattering of the concrete). These models were applied successfully in the FEMA 439A study of the reinforced concrete Murrah Building.

Significantly less field testing of structural steel elements and frame systems subjected to blast loading has occurred, so SDOF models cannot be relied upon to the same extent that they can be for reinforced concrete elements. The analyses described in Sections 3.3 through 3.7 illustrate the need for further field testing of a number of key aspects of steel frame systems.

At the localized, elemental level, better understanding of buckling characteristics of element webs and flanges; flexural-torsional response modes; beam-column joint behavior; and column splice performance are needed. The existence of bolted, riveted (older buildings), and welded connection details further complicates analysis, because of the variety of details that can be used to create a connection between two steel elements. It is likely that these significantly different connection techniques provide significant differences in blast response. These differences are exacerbated by variations in connection techniques used in different generations of steel design and different material characteristics used in each generation.

While such an approach was avoided in this study, it is also noteworthy that many analysts, for lack of any better information, often use results of seismic testing as the basis for blast response modeling. The typically cyclic nature of seismic testing is significantly different from the highly impulsive, and generally monotonic, blast-loading environment. Such assumptions may not necessarily be appropriate and could lead to inaccuracies in themselves.

In addition to localized element response issues, modeling global response using SDOF models is complicated because such models consider only individual beams or columns with assumed support conditions. To consider system-wide response, the analyst must somehow "sum" the responses of the individual elements outside the SDOF models.

Often in blast design of new structures and evaluation of existing structures, clients are willing to pay for design conservatism, which permits the designers and analysts to use worst-case scenario results generally used in this study. This, however, does not lead directly to accurate understanding of response.

The nonlinear FEA results closely paralleled the results of the limited field testing that has been accomplished in connection with this study. It is noteworthy that the analyses were completed before the testing was conducted. At present, such analyses are very complex to set up, very computationally intensive to run, and require analysts with considerable expertise to interpret the results. Only a limited number of engineers who are involved with blast-related research are currently equipped to perform these analyses, and typical design projects will not support the costs associated with them.

There are also still significant gaps in the ability of FEA to predict several potential failure modes accurately without further test results to calibrate the analysis. This is particularly true when the blast loads are intense enough to cause a fracture within the steel element and when complex connections are impacted – the field tests reported herein did not incorporate column splices or beam-column connections.

The shortcomings of the SDOF modeling highlight a significant need for additional testing, paralleled by detailed nonlinear FEA, to gain better understanding of blast response of steel structures.

3.11 Blast Analysis Summary

Table 3-1 summarizes the blast responses of key elements in the structure, based on the results of the SDOF model results, the FEA, the blast test, and the engineering judgment of the project team. Members indicated as "Lost" were judged to have been damaged sufficiently to be ineffective in supporting any gravity load following the blast. Members that are indicated as "Damaged, but intact" refer to elements that sustained blast-induced permanent damage, but not enough to render them incapable of supporting some gravity load. Figures 3-19 through 3-22 show elevations of the G-Line frame for each structure indicating the members that were assumed to be lost, damaged, and intact. The adequacy of those members to support the estimated gravity loads on the structure after the blast is discussed in Chapter 4.

Table 3-1. Blast Response of Key Elements in Structure

	Original Structure	Connection Upgrade Scheme	Exterior BRBF Scheme	Interior BRBF Scheme	Re-detailed Frame
Column G3	Lost Due to Splice Failure	Damaged, but intact	Damaged, but intact	Damaged, but intact	Damaged, but intact
Girders 2nd Floor	Lost Between G2 and G4	Lost Between G2 and G4	Lost Between G2 and G4	Lost Between G2 and G4	Lost Between G1 and G4
Girders 3rd Floor	Lost Between G2 and G4	Damaged, but intact	Damaged, but intact	Lost Between G2 and G4	Lost Between G2 and G4
Floor Slab	Lost Between G1-G5 and F1-F5	Lost Between G1-G5 and F1-F5	Lost Between G1-G5 and F1-F5	Lost Between G1-G5 and F1-F5	Lost Between G1-G5 and F1-F5

3.12 Figures for Chapter 3

Figure 3-1. Explosive Threat Similar to the Oklahoma City Murrah Building Bombing

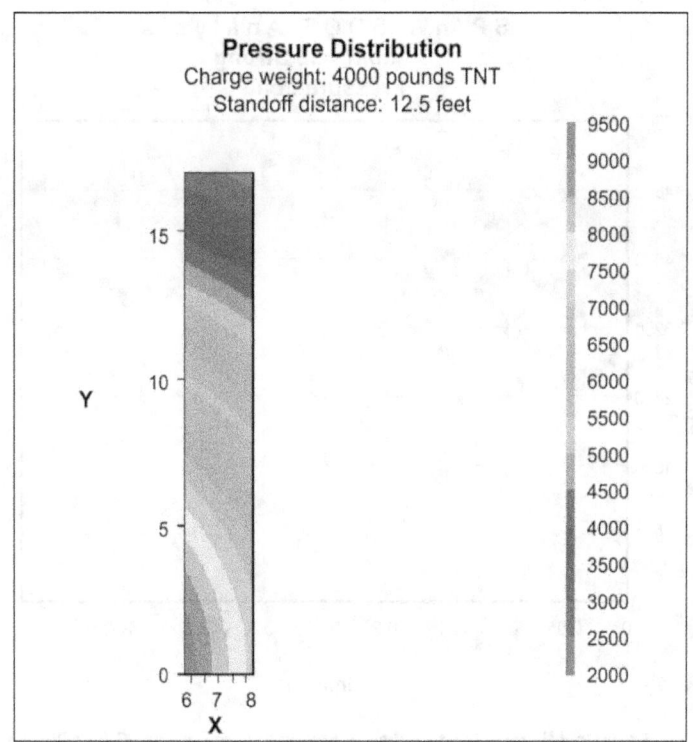

Figure 3-2a. Reflected Pressure on Front Face of Column G1 (First Floor)

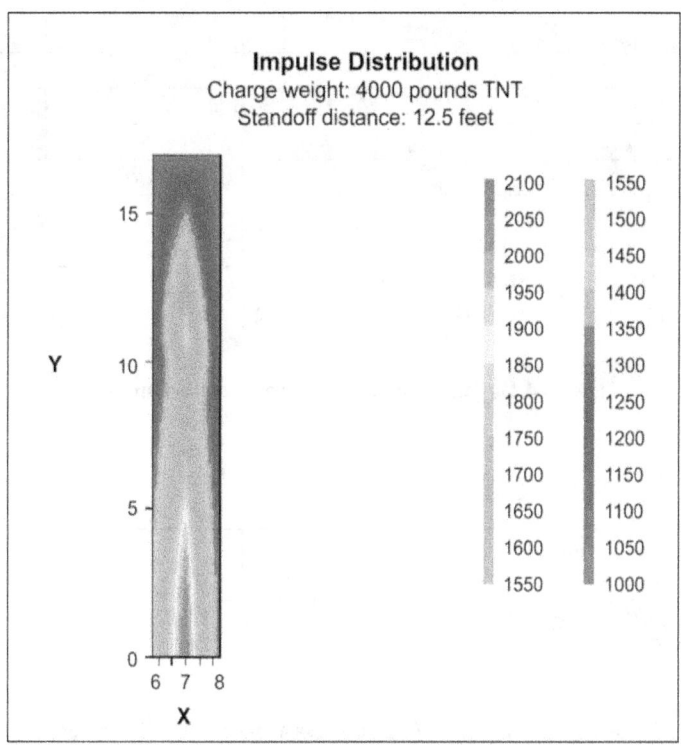

Figure 3-2b. Reflected Impulse on Front Face of Column G1 (First Floor)

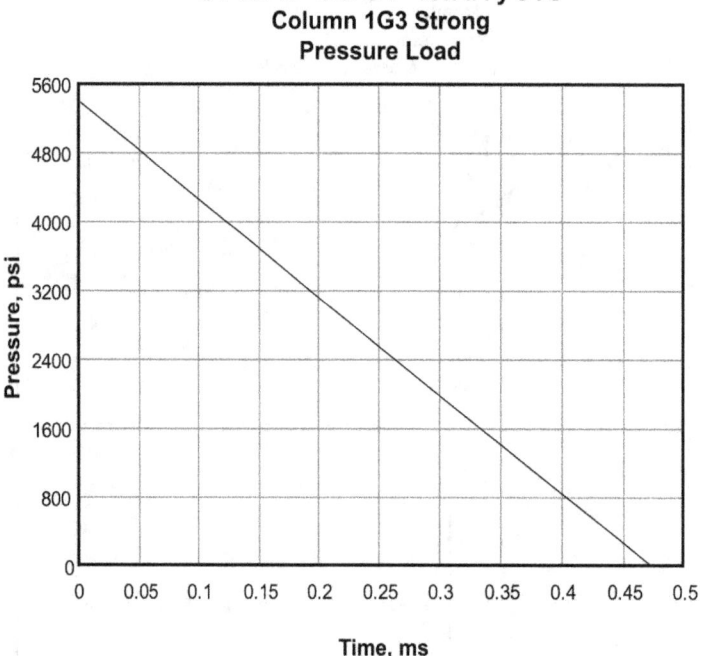

Figure 3-3. Equivalent Uniform Load Applied to Strong Axis of Column G3 for SDOF Analysis

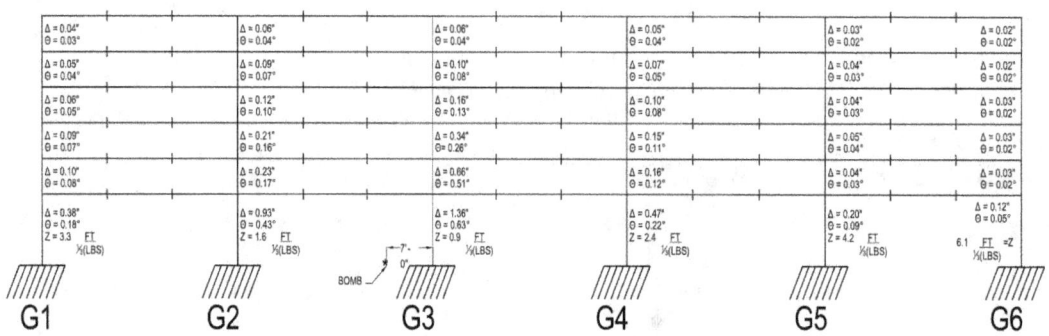

Figure 3-4. Maximum Strong Axis Response on Column Line G

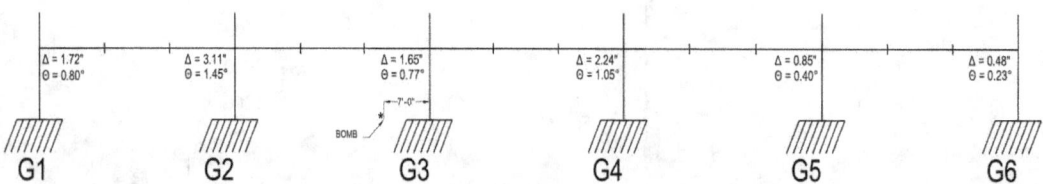

Figure 3-5. Maximum Weak Axis Response on Column Line G (First Story)

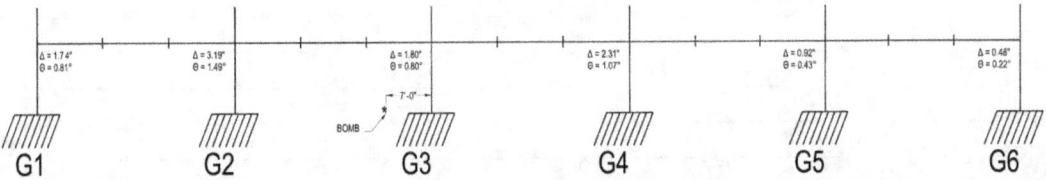

Figure 3-6. Maximum Biaxial Response on Column Line G (First Story)

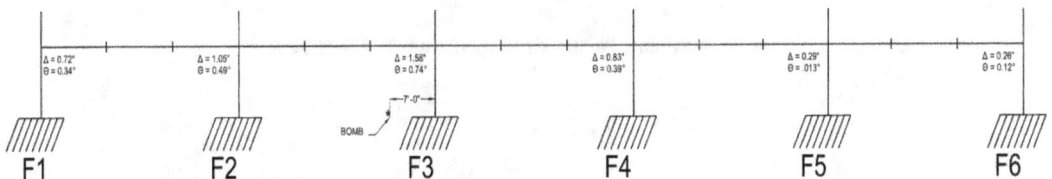

Figure 3-7. Maximum Biaxial Response on Column Line F (First Story)

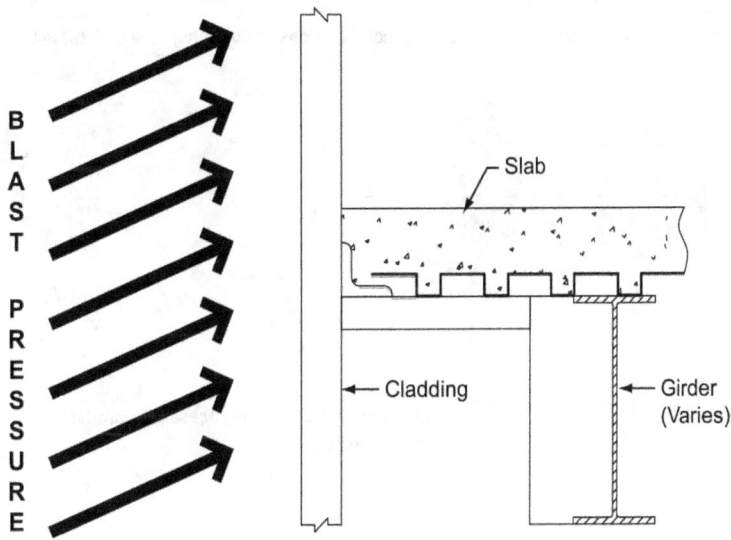

Figure 3-8. Blast at Girder / Cladding

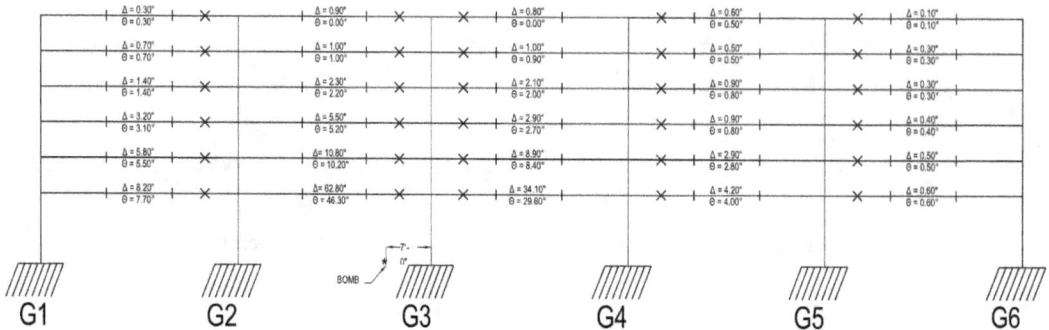

Figure 3-9. Maximum Weak Axis Response on West Side Girders

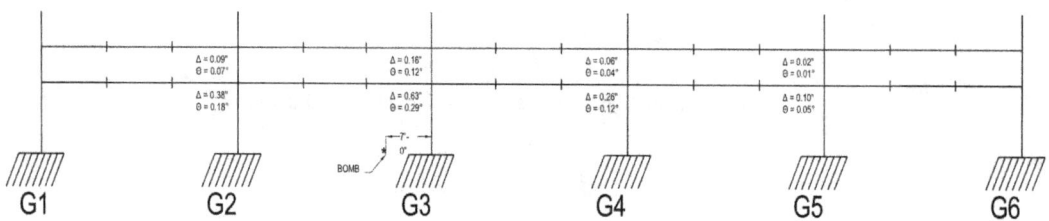

Figure 3-10. Maximum Strong Axis Response on Concrete-Encased Columns; Line G (First & Second Stories)

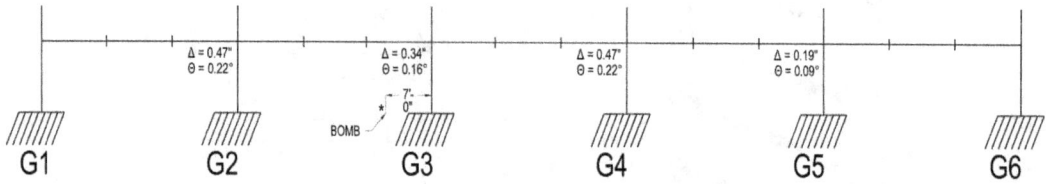

Figure 3-11. Maximum Weak Axis Response on Concrete-Encased Columns; Line G
(First Story)

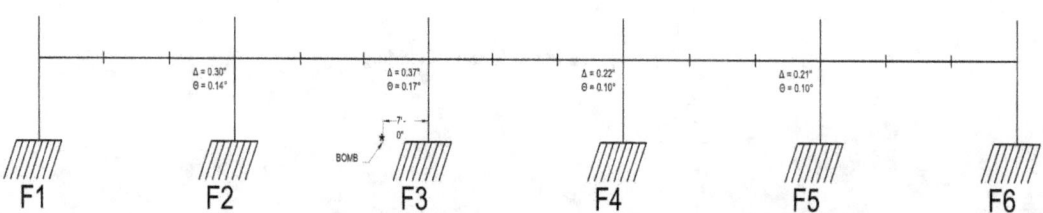

Figure 3-12. Maximum Weak Axis Response on Concrete-Encased Columns; Line F (First Story)

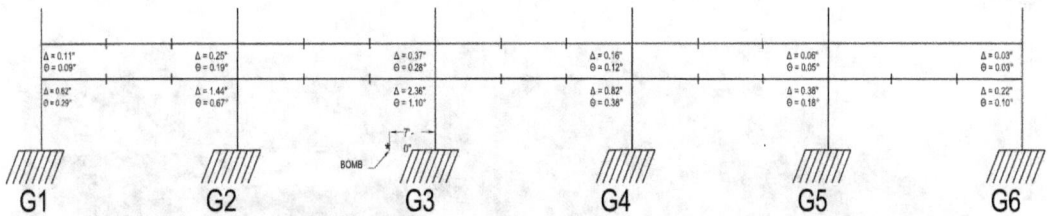

Figure. 3-13. Maximum Weak Axis Response of Re-detailed Columns; Line G
(First & Second Stories)

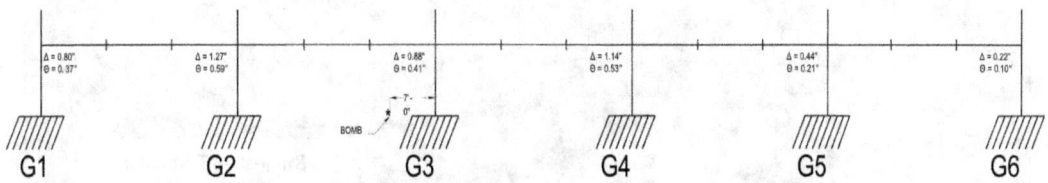

Figure. 3-14. Maximum Strong Axis Response of Re-detailed Columns; Line G (First Story)

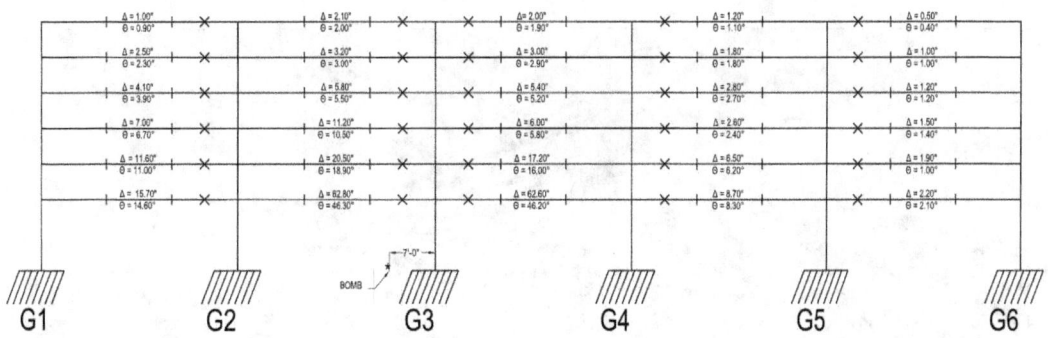

Figure 3-15. Maximum Weak Axis Response; West Side Girders with "Dogbone" Connections

Figure 3-16. Test Column and Explosive Charge

Figure 3-17. Post-Blast Condition of Column

Figure 3-18. Post-Blast Condition of Column (Close-up)

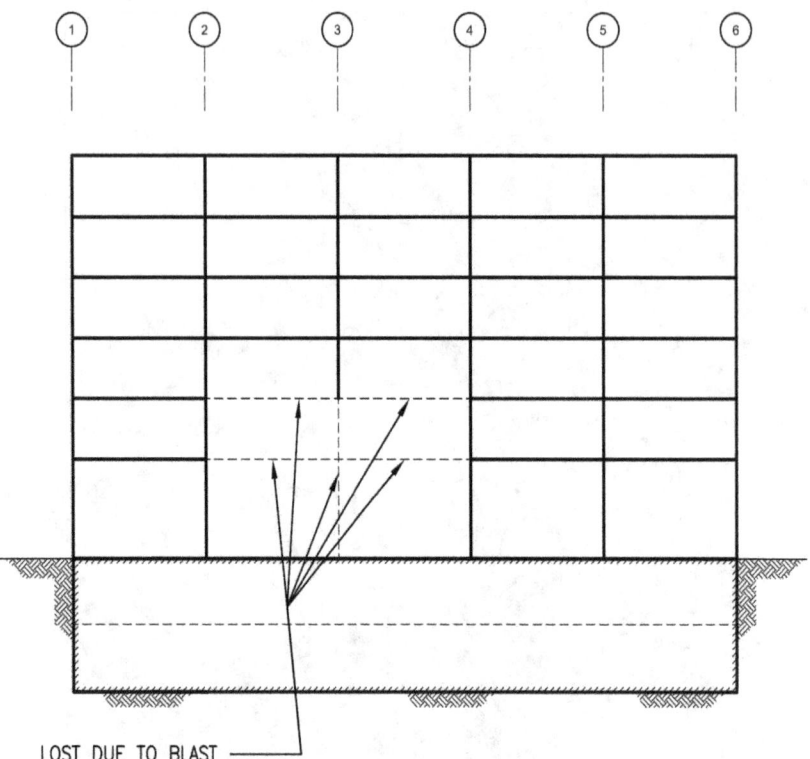

LOST DUE TO BLAST

Figure 3-19. Original Building and Interior BRBF Upgrade Post-Blast

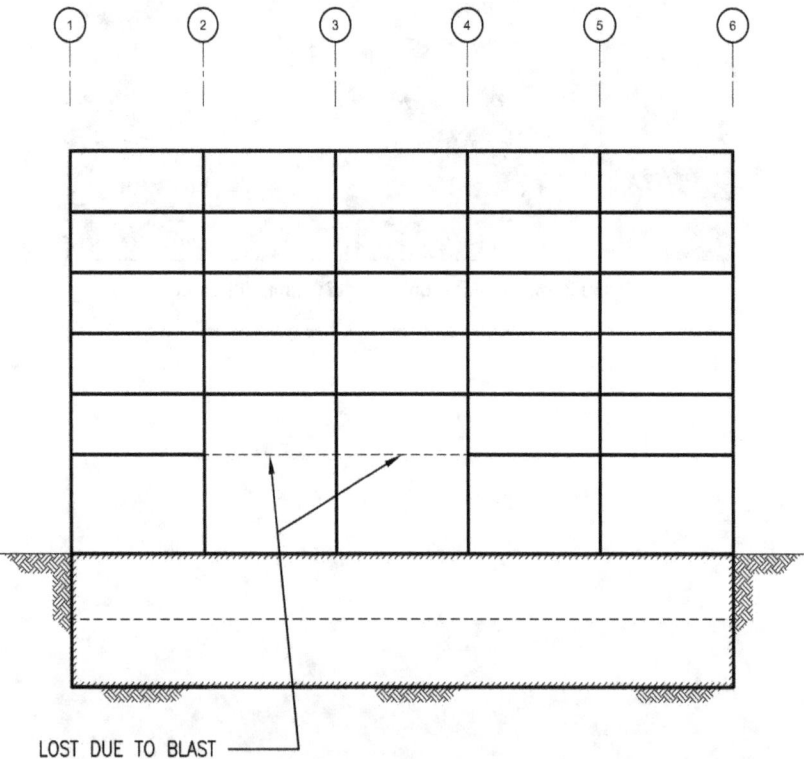

LOST DUE TO BLAST

Figure 3-20. Moment Frame Connection Upgrade Post-Blast

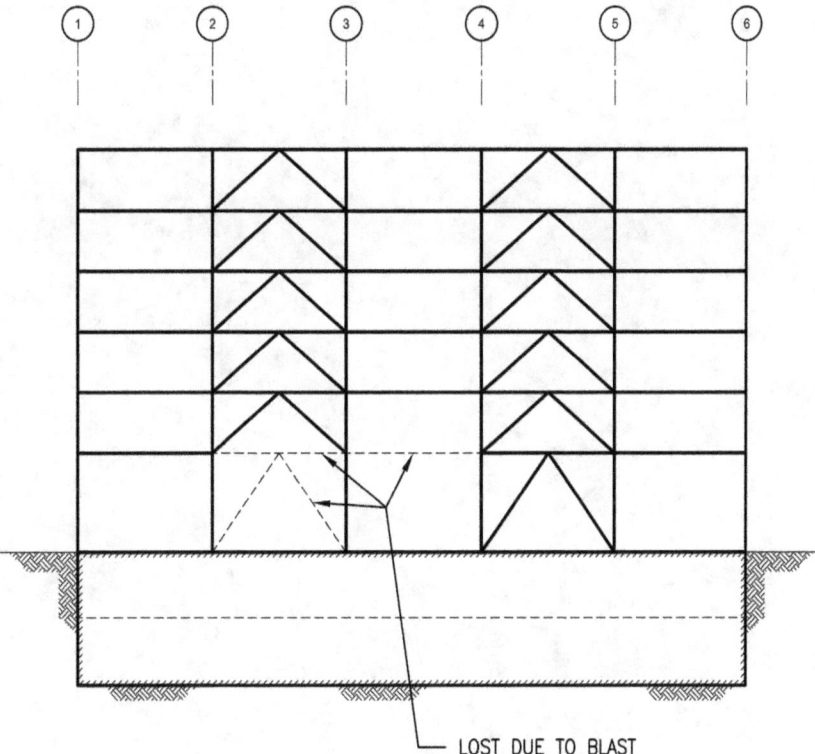

Figure 3-21. Exterior BRBF Upgrade Post-Blast

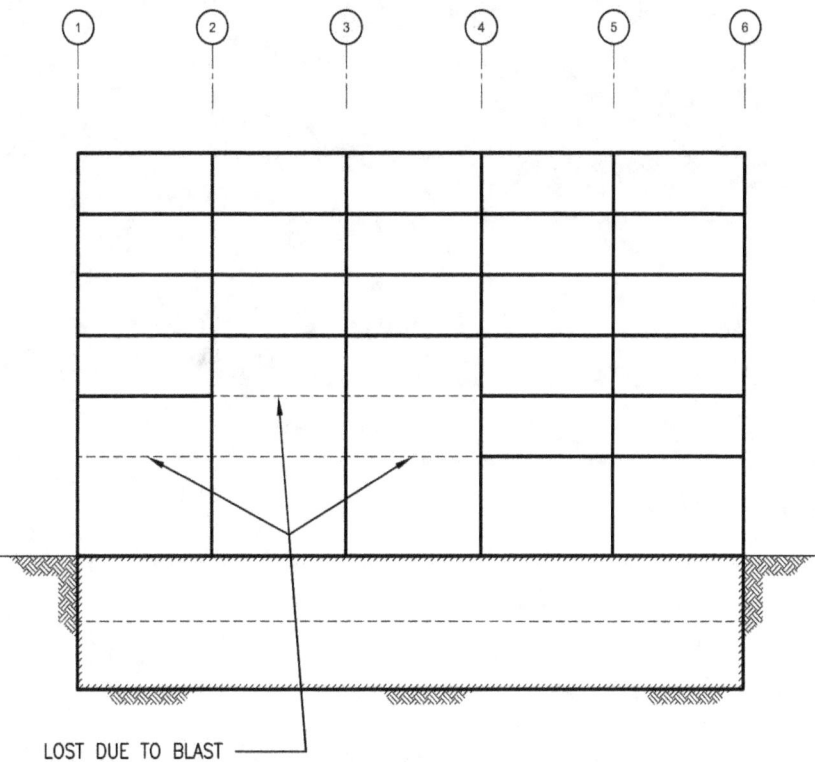

Figure 3-22. SMRF Re-detailed Building Post-Blast

4 Progressive Collapse Evaluation of Frames Strengthened to Improve Earthquake Resistance

4.1 Introduction

The final phase of this study involved assessing likely progressive collapse scenarios, following the blast event described in Chapter 3. Scenarios were postulated for the original structure, each of the proposed seismic strengthening schemes, and the re-detailed exterior frame. To reiterate a previous comment, unlike the Murrah Building study reported in FEMA 439A, no actual event has produced a baseline for calibrating the analyses. There are many uncertainties with the blast analyses described in Chapter 3. Those uncertainties impacted the collapse evaluation and were compounded with significant uncertainties that exist with respect to the behavior of steel frame structures at their collapse state. There is therefore a wide range of potential outcomes for the original building's response to the blast. Because of the many uncertainties, worst-case and best-case post-blast collapse scenarios are discussed.

4.2 Approach

The progressive collapse assessments involved the specialized gravity-load structural analyses of each system. The analyses combined a series of basic hand calculations and linear structural analyses with engineering evaluations of building response. For each scenario, members representing those estimated to be destroyed by blast load effects were removed from the system analytical model. Calculations were then performed to determine if the "blast-damaged" structure could support its gravity loads. Gravity loads are assumed to be the building's self-weight – 75 pounds per square feet (psf) for the dead load of the floors and roof, and 45 psf for the cladding load of the surface area at each story level, plus 25% of the estimated 40 psf live load for office space and 20 psf roof live load. If frame capacity was insufficient to resist these gravity loads elastically, then virtual work analyses were performed to determine if the post-elastic frame capacity was sufficient after the structure yielded and redistributed forces.

Since blast-damaged beams and columns are removed almost instantaneously in actual blast events, loading immediately following the blast experienced by the structure will be increased above the nominal gravity-load by impact effects. Typically recommended progressive collapse analysis procedures, such as the GSA and DoD Guidelines (References 33 and 40, respectively) conservatively double applied gravity loads to account for impact effects. Using this rationale, the surviving structural elements must be twice as strong as the calculated gravity load demands to prevent collapse. In this study, capacity-to-demand (C/D) ratios were calculated for each element in each structural frame system that was assessed. Each system consisted of those elements that remained after blast damage occurred, following removal of those members that were assumed to be destroyed in the blast.

With the assumed doubling of gravity-load forces to account for impact effects, a calculated system C/D ratio of 2.0 or larger for the unfactored (i.e., no load factors, commonly used in design, were applied) gravity loads implies that a progressive collapse mechanism will not form. If the C/D ratio for the unfactored gravity load was less than 1.0, progressive collapse was predicted to be certain.

If the C/D ratio was more than 1.0 but less than 2.0, collapse was deemed possible. The potential failure mechanism was then examined more closely before postulating that a collapse mechanism would occur. If the failure mechanism was brittle (e.g., connection failure), then collapse was assumed to occur. If the failure mechanism was ductile (e.g., flexural yielding of beams), then, using a virtual work approach, an energy method solution was performed to estimate the amount of post-yield displacement the frame would undergo before arresting the collapse. The energy method is shown graphically in Figure 4-1.

A load-displacement "capacity curve" of the frame's capacity at the critical location was constructed using, first, the elastic analysis results to establish initial stiffness and yield and then virtual work evaluation to develop post-yield maximum vertical force capacity, assuming ductile yielding in the beam flanges. The connections were initially assumed to be robust enough to permit yielding in the beam flanges.

After the capacity curve was developed, connections were then subjectively evaluated to determine if they could accommodate calculated support rotations under the applied loading "demand curve." A constant force representing the gravity load was superimposed over the capacity curve. Downward column displacement was found at the point where the area under the "capacity curve" equaled the area under the gravity load, or "demand," curve. The displacement was an estimate of maximum displacement at the location being examined, including the increase for impact, that the frame would undergo after members are lost due to direct blast effects, as the frame attempts to carry the redistributed gravity loads, or to "resist" collapse. The final displacement was compared to the yield displacement to determine ductility demand on the frame and compute support rotations at all connections required to accommodate the maximum displacement. Comparison of rotations calculated from the displacement to estimated rotation capacities establishes the potential for the frame to resist collapse.

A large source of uncertainty in both the blast and progressive collapse analyses is the final condition of the girders and their residual flexural capacities in that condition. The SDOF models are likely conservative in their prediction of weak-axis displacement, because restraint due to slab effects was ignored. The FEA model included slab effects, but only in adding rigid restraints to beam top flanges. Neither analysis considered brittle failure of connection welds; only very limited experimental data on such failures are available. Therefore, the post-blast conditions of the girders are uncertain. Even if the conditions were known with more certainty, computing reduced girder capacities due to possible out-of-plane deformations of their bottom flanges and webs is difficult without localized nonlinear FEA, which are outside the scope of this study.

4.3   Performance of the Original Building

In the original building, Column G3 was assumed to have been severed at the second story splice, and the second and third floor girders were assumed to have lost their connections to the columns. The remaining beams framing into Column G3 could be significantly damaged or distorted.

Collapse potential was first assessed for the frame assuming that the framing beams could yield. From the virtual work evaluation, it was found that the C/D ratio was 1.5, indicating that the frame had the capacity to resist gravity load (since C/D > 1), but that the frame would yield due to dynamic impact effects (since C/D <2). The energy method evaluation predicted the maximum displacement of column line G3 to be 6 inches downward, requiring a support rotation of 0.017 radians, with a corresponding displacement ductility ratio of 1.3 for the beam hinges. This implies that, if the girders could yield without connection failure (due either to weld failures or premature buckling due to distortion from the blast), then the frame would arrest the collapse because the rotational and displacement ductility demands are not very large.

The blast analysis indicated that the second floor slab one bay deep into the building, between Columns G1 and G5, would be lost directly due to blast. If collapse does not occur, the fourth, fifth, and sixth floors, and the roof areas would likely survive. The condition of the third floor area then becomes critical to the assessment, since the lower floor girders between Columns G2 and G4 were assumed to have been blown away from their connections by the blast. The G Column Line ("G-Line") girders have secondary beams framing into them. If those girders fail, then half the floor support is lost. It is uncertain whether the shear tab connections attaching the girder webs to the columns would fail due to the blast in addition to the beam flange welds failing. If the shear tabs are not lost, the girders would be capable of supporting the floor beams as simply supported beams. If the girders cannot support the floor beams because the shear tabs are lost, then the slabs would drape between column lines 2, 3, and 4, potentially forming a catenary-type resistance mechanism. It is uncertain whether this would be possible in the prototype frame, given the lack of information on the welding for the existing metal deck and splicing information for the welded-wire-fabric in the concrete topping slab. In past instances, particularly 130 Liberty Plaza following the September 11, 2001 terrorist attacks on the nearby World Trade Center, this type of flooring was observed to have formed a catenary-type mechanism and resisted collapse, albeit with very large displacement (Reference 37).

Because of the many uncertainties involved, the final condition of the building after the blast cannot be established definitively. The best-case scenario is that the second floor area one bay in from the building face between Columns G1 and G5 (approximately 3,600 sf) would be lost, but without the collapse of the remaining floors– as shown in plan view in Figure 4-2. The worst-case scenario is that all of the floor area between Columns G2 and G4 (12,600 sf) would be lost in addition to the second floor area lost due to the direct blast.

4.4   Performance of the Moment Frame Connection Upgrade Scheme

In the Moment Frame Connection Upgrade Scheme blast analysis, it was concluded that Column G3 would survive the direct blast effects (because of the column splice upgrade), the second floor slab one bay deep into the building between Columns G2 and G4 would be lost, and the second and third floor girders on the building face would be lost. Since Column G3 was not lost, there would be no large-scale collapse. Depending on the ability of either the third floor girder's shear tabs to survive the blast intact or the slabs to support catenary action between column lines, either the second floor between Columns G1 and G5 one bay inside the building (3,600 sf) or both the second and third floors (5,400 sf) would be lost. A new C/D analysis was not required, because appropriate observations could be made using the original building performance analysis.

4.5   Performance of the Exterior BRBF Scheme

The blast analysis of the exterior BRBF scheme showed that Column G3 would not be lost and that girder and slab damage would be the same as the other schemes. Because parts of the girder ends are within the concrete encasement that is a part of this scheme, there is greater potential for the third floor beam to remain intact and therefore support that floor than was observed for the Moment Frame Connection Upgrade Scheme. For this scheme, it was therefore concluded that only the second floor slab between Columns G1 and G5 one bay inside the building (3,600 sf) would be lost.

4.6 Performance of the Interior BRBF Scheme

Since the G-Line frame in the Interior BRBF Scheme was not affected, the performance of this scheme is anticipated to be the same as the original building.

4.7 Performance of the Re-detailed Frame

The behavior of the re-detailed frame is very similar to that of the Moment Frame Connection Upgrade Scheme. The only substantial difference is the loss of an additional floor beam. No additional C/D analysis is required to assess performance. As with the Connection Upgrade scheme, it is possible that the third floor girders' shear tabs remain intact or that the floor supports itself through catenary action. Depending on the third floor response, either the second floor between Columns G1 and G5 one bay inside the building (3,600 sf), or three floors between those columns (7,200 sf) would be lost.

4.8 Collapse Evaluation Summary

Table 4-1 summarizes the worst-case assessment of the post-blast collapse of the structure, and Table 4-2 summarizes the collapse assessment of the structure assuming the best-case scenario. For all the structures, the best-case scenario indicates that the only floor area that would be lost is that due to direct blast damage, and the steel framing would be capable of arresting any collapse following the blast.

In the worst-case scenario, every scheme has some post-blast collapse, except the scheme specifically designed to consider arresting a perimeter collapse – the Exterior BRBF with Hat Truss. The original structure experiences the largest collapse area, between lines G2 and G4 one bay into the building.

Table 4-1. Summary of Estimated Blast and Progressive Collapse Damage for Worst-Case Scenarios

Floor Level	Floor Area (sf)	Direct Blast Damage Only (sf)	Total Floor Area Lost				
			Original Building (sf)	Interior BRBF Scheme (sf)	Connection Upgrade Scheme (sf)	Exterior BRB + Hat Truss Scheme (sf)	Re-detailed Perimeter Frames (sf)
Roof	27,000	0	1,800	1,800	0	0	0
6th	27,000	0	1,800	1,800	0	0	0
5th	27,000	0	1,800	1,800	0	0	0
4th	27,000	0	1,800	1,800	0	0	0
3rd	27,000	0	1,800	1,800	1,800	0	1,800
2nd	27,000	3,600	3,600	3,600	3,600	3,600	3,600
Total Above 1st Floor	162,000	3,600	12,600	12,600	5,400	3,600	5,400
% of Total Floor Area Damaged		2%	8%	8%	3%	2%	3%
% of Damaged Area Due to Blast		-	29%	29%	67%	100%	50%
% of Damaged Area Due to Progressive Collapse		-	71%	71%	33%	0%	50%

Table 4-2. Summary of Estimated Blast and Progressive Collapse Damage for Best-Case Scenarios

Floor Level	Floor Area (sf)	Direct Blast Damage Only (sf)	Total Floor Area Lost				
			Original Building (sf)	Interior BRBF Scheme (sf)	Connection Upgrade Scheme (sf)	Exterior BRB + Hat Truss Scheme (sf)	Re-detailed Perimeter Frames (sf)
Roof	27,000	0	0	0	0	0	0
6th	27,000	0	0	0	0	0	0
5th	27,000	0	0	0	0	0	0
4th	27,000	0	0	0	0	0	0
3rd	27,000	0	0	0	0	0	0
2nd	27,000	3,600	3,600	3,600	3,600	3,600	3,600
Total Above 1st Floor	162,000	3,600	3,600	3,600	3,600	3,600	3,600
% of Total Floor Area Damaged	2%	2%	2%	2%	2%	2%	
% of Damaged Area Due to Blast	-	100%	100%	100%	100%	100%	
% of Damaged Area Due to Progressive Collapse	-	0%	0%	0%	0%	0%	

4.9   Figures for Chapter 4

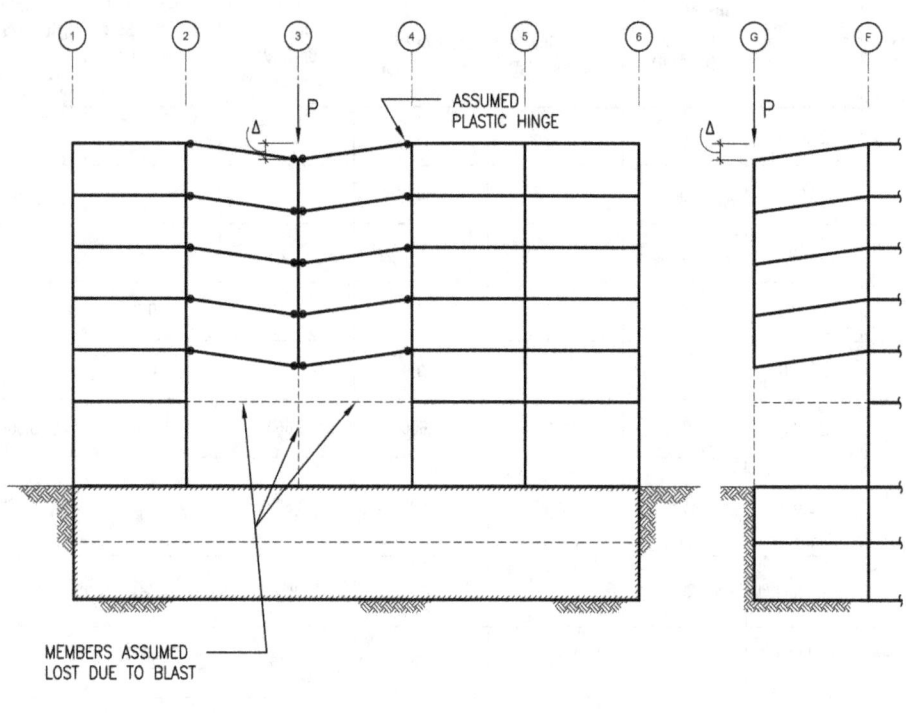

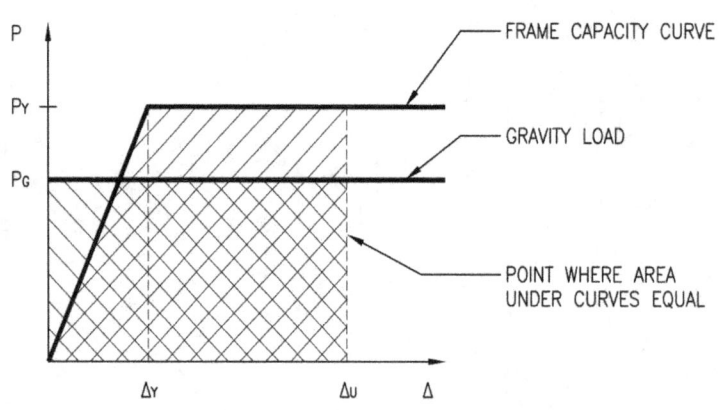

LEGEND

P$_Y$ FRAME YIELD STRENGTH

P$_G$ GRAVITY LOAD ON FRAME

Δ$_Y$ FRAME CONTROL NODE YIELD DISPLACEMENT

Δ$_U$ FRAME CONTROL NODE ULTIMATE DISPLACEMENT

Figure 4-1. Energy Method Post-Yield Displacement of Frames

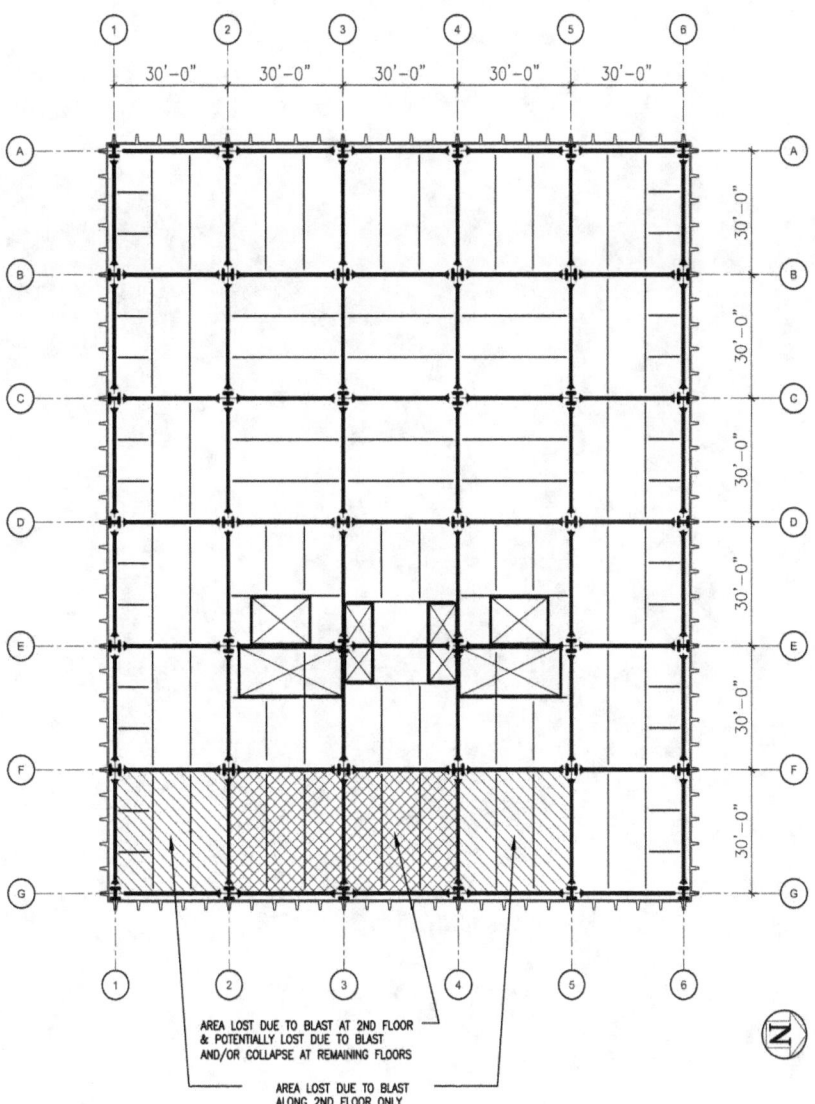

Figure 4-2. Original Building Best-Case Scenario Post-Blast

5   Conclusions and Recommendations

5.1   Overview

This study has examined the relative effectiveness in improving blast and progressive collapse resistance for older structural steel moment frames not detailed for seismic resistance that might be achieved by applying seismic upgrade measures that are associated with current-day structural engineering practice in areas of high seismicity. The study has been largely analytical in nature, supplemented by a blast and axial load testing of a column similar to the column most severely exposed to the simulated blast effects in the numerical analyses for this project. The scenario of the 1995 Murrah Federal Building bombing was employed as a benchmark for the study. Numerical analyses were performed on the "original" structure that had no structural modifications, on the three upgraded structural systems consisting of an Interior Buckling-Restrained Braced Frame (BRBF) scheme, a basic Connection Upgrade scheme, and an Exterior BRBF with an added Hat Truss, and the original structural configuration that had been re-detailed using modern building code requirements for structural steel moment frames in regions of high seismicity.

In Chapter 3, the blast analysis, supplemented with nonlinear finite element analysis and limited field testing indicated that the steel columns exhibited substantial toughness in their response to the blast. The beams, however, appeared to be more susceptible to blast-induced damage. This was due in part to their larger surface area normal to the blast and the lower lateral stability of typical beam sections.

The response of the beams was one of the two largest sources of uncertainty in the blast and collapse analyses. The other major source of uncertainty was the performance of the connections, both the column splices and the beam-column connections. Because of these two uncertainties, it was felt that providing two scenarios, a "best case" and "worst case," for the progressive collapse evaluation would be appropriate.

The progressive collapse analyses described in Chapter 4, which were performed following the blast response analyses described in Chapter 3, concluded that even in the "baseline" structure, which was considered to be "nonductile" and seismically vulnerable, the floor area that would be lost would range from 2% to 8%, a relatively small percentage of the total. The three seismic strengthening schemes and the re-detailed scheme improved the worst-case projection by varying amounts. Table 5-1 presents the best-case and worst-case estimates of the floor areas lost due to the bombing.

Table 5-1. Floor Area Loss Summary

	Total Floor Area Above First Floor (sf)	Direct Blast Damage Only (sf)	Total Floor Area Lost				
			Original Building (sf)	Interior BRBF Scheme (sf)	Connection Upgrade Scheme (sf)	Exterior BRB + Hat Truss Scheme (sf)	Re-detailed Perimeter Frames (sf)
Best-Case	162,000	3,600	3,600	3,600	3,600	3,600	3,600
Worst-Case	162,000	3,600	12,600	12,600	5,400	3,600	5,400
Worst-Case % Change			Baseline	No Change	60% Less	70% Less	60% Less

At first, the relative improvements afforded by the Moment Frame Connection Upgrade Scheme, the Exterior Buckling Restrained Braced Frame – BRBF, and the re-detailed perimeter frame appeared to provide minimally improved performance. It appeared that way because the area of the "original" building lost was low and because the best case estimates indicated that the structure would survive. However, when the percent change in area lost for the worst case was examined, one could observe that there was sub-stantial improvement in the blast resistance of the structure due to the exterior seismic upgrade designs and the seismic detailing.

The Connection Upgrade Scheme largely improved survivability through improving the blast resistance of Column G3 with its column splice upgrade. The Interior BRBF did not strengthen the exterior column line, so its survivability would largely match the original building's survivability. The re-detailed perimeter frame performed very simi-larly to the Connection Upgrade Scheme.

The Exterior BRBF with a hat truss provided slightly improved performance (2% lost floor area for both best and worst cases). The chief contributor to its improved perfor-mance was the concrete encasement of columns (and corresponding floor beam ends) in the exterior column lines, which was included in that scheme as a means of increasing column axial capacity.

5.2 Conclusions

A review of the study points to several significant observations that can be made about the effects on survivability that seismic upgrade designs, building configuration, past and present construction practices, and seismic detailing provisions provide. While there are many significant unknowns that require further study for improving the understanding of blast and progressive collapse responses in steel moment frames, some conclusions can be drawn.

The main conclusions from the study are listed below. The conclusions apply specifically to the subject building, explosive weight and standoff distance.

- The subject building performed well, having a low percentage of floor area lost under blast and progressive collapse.

- Regularity, redundancy, and configuration led to improved performance in response to the blast.

- The blast and progressive collapse resistance performance of a building subjected to an exterior detonation can be greatly improved by the use of seismic upgrade schemes along the perimeter of the building.

- Improved blast and collapse performance resulted from column splices that provided substantial flexural and shear continuity between the columns and column base details where the base plates were sufficiently embedded in concrete along the perimeter frame.

- To maximize the blast resistant benefits of a seismic upgrade design, one must take into account multi-hazard mitigation principles relating to where the new seismic upgrade elements should be located and what elements should be strengthened.

- Seismic upgrade of interior structural elements did not provide any significant increase in blast and progressive collapse resistance performance.

- SDOF analyses were found to be accurate in this specific example when the scaled range was greater than approximately 1.0 lbs/ft$^{0.33}$, the pressure distribution along the height of the member was somewhat uniform, and the shockwave arrived at the face of the element at the same instant along its length.

- Based on the analytical models and the physical test, it appeared that the steel column in this building performed well under an explosive scenario employing a very large charge situated a moderate distance away.

- Specialized Finite Element Analyses (FEA) used in this evaluation predicted the deformation of the tested column with reasonable accuracy.

- There were significant unknowns in the characterization of the blast demands and the blast response of steel connections (beam-column joints, column splices, column bases) that impacted the response estimates.

5.2.1 Structural Configuration Impact on Response

It is noteworthy that the best-case blast/collapse scenario found that only a small portion of the second floor slab was lost due to air blast, with no ensuing collapse. Thus, the best-case assessment suggests that a regularly configured, redundant (due to the presence of all the moment connections) and well-detailed structure potentially has substantial toughness to prevent a disproportionate loss of floor area in a blast event.

In contrast with the lack of redundancy and the variety of plan irregularities in gravity load elements and paths in the Murrah Building (FEMA 439A), there was significant redundancy and a high degree of plan regularity in the building studied here. This building footprint was comprised of six column lines in one major dimension and seven column lines in the other major dimension. Column spacing was a uniform 30 feet for all column lines, throughout the building height. Because the steel building was six bays deep normal to the face where the vehicle bomb was located, even if all five bays along that face were lost, only 17% of the total floor area would be lost.

Conversely, the Murrah Building's primary framing system contained only three column lines through the depth of the building, and the two lower floors had numerous discontinuous gravity load paths, including the doubling of column spacing relative to the upper stories in the long axis dimension of the building. In the Murrah Building, with it being only two bays deep, the loss of the street face bays led to a loss of about half the floor area.

Therefore, from the standpoint of blast resistance, it appears that a plan configuration that lends itself to multiple frame bays in each direction may provide better overall performance than one that has a high aspect ratio with only a few bays in one direction.

5.2.2 Benefits of Seismic Upgrade Designs

Although the expected loss of floor area, even in the worst-case scenarios, was relatively small, as noted above, there did appear to be some significant gains from seismic upgrade, depending on configuration. The Connection Upgrade Scheme resulted in 57% reduction in lost floor area. The Interior BRBF scheme did not reduce loss because no modifications were made to the perimeter frames. The biggest improvement, the Exterior BRBF with Hat-Truss, results in a reduction of 71% in the floor area lost. These improvements were for the worst-case scenario, because in the best-case scenario the original structure and all the upgraded ones resist a post-blast collapse.

The greatest benefit in the Exterior BRBF scheme came from the concrete encasement of exterior columns. By increasing both column mass and section capacity, the blast resistance increased substantially. It is also postulated that the concrete encasement would protect the existing beam-column connections. Upgrading the column splice to permit the column to yield in a flexural manner also helped the blast resistance because the G3 column was not lost.

In this particular example, the hat truss, added solely to mitigate the potential loss of a column, had little impact on the structure because it was determined that no columns were lost due to blast. However, had that not been the case, the hat truss would have bridged over the lost column.

5.2.3 Contrast of Past and Present Construction Practices

While the lack of testing and analysis of steel moment frames like that studied here limit detailed conclusions, several differences between 1970's era construction practice and current practice could possibly impact blast response.

First, virtually every beam-column connection in the original building considered in the study was a full moment-resisting connection. This was common practice at that time and may have been done to reduce the gravity load moments on the beams, permitting lighter sections to be used. This may have been done because the cost of additional steel would have exceeded the added labor cost to construct those connections. Today, the opposite is generally true - the labor associated with a full moment-resisting connection is great enough to steer designers toward using as few moment frame lines with as few bays as possible.

Regardless of the Seismic Design Category of the structure, if it were designed and built today, it is likely there would only be discrete moment frames, of two or so bays long in any given column line. While it is likely that the moment frames would be located along the exterior of a building today, there is no guarantee of that. Beams would most likely not frame normal to the exterior frame with moment connections to provide added resistance, as they do in the building in this study.

Second, current design practice for moment frames typically ensures that the exterior column lines in the building would have columns oriented so that their strong axes would be in the plane of the frame. This would permit the columns to provide maximum flexural strength and stiffness in the plane of the column line (corner columns would probably be oriented so that their strong axes would be in the plane of the shorter column line). It is also possible that the column sections chosen could have a higher aspect ratio than W14 columns, such as W24. W24 sections are sometimes used in lieu of W14 sections to gain additional stiffness and strength for the moment frame columns for less weight than W14 sections due to their higher aspect ratio in the plane of the frame. However, their use results in columns with thinner flanges and webs that can be more susceptible to local buckling failures and have a larger surface area that can be loaded with the air blast or cladding debris.

In this building, the W14 column studied was oriented so that its strong axis was perpendicular to exterior frame line. This would permit the column to resist external blast or impact loads with strong axis-dominated bending, whereas current practice would likely force the column to resist external blast or impact loads in weak axis bending, possibly with a lower weak-axis moment of inertia and larger loaded area.

Both the column size and orientation differences may well result in the older design's being less vulnerable to external blast loading than a more recently designed structure.

Possibly offsetting the potentially lesser vulnerability in the older structure because of column section properties and number of moment connections, the older connections may not be as robust as current connections and were made using welds made with less quality control than current connection details require, especially in moderate or high Seismic Design Categories. In the column splices that were located just above the second floor slab, the older connections had flange-only welds, whereas current practice would incorporate both flange and web welds. Today, the columns splices would also be located slightly higher above the floor than they were in this building.

The column-to-girder connections in the original structure had beam flanges directly welded to the column and bolted webs. These connections, common in older steel-frame buildings, may lessen structural integrity relative to current practice due to the lack of toughness of some of the older beam flange weld material, the lack of weld quality control provisions that are now required, and many older full-penetration weld geometries being more fracture sensitive than current ones. The connections were found to experience fractures initiating in the beam flange welds in the 1994 Northridge earthquake.

However, the severity of these older connections on the blast response is uncertain for several reasons. First, the seismic response differs from the response of a connection to a blast. In the seismic response, the connections undergo reversals of load, causing low-cycle fatigue issues because of the fracture-sensitive details. In a blast response, generally, the connection will be loaded in one direction without load reversals. Second, the lesser weld quality control and the unknown toughness of older welds does not necessarily guarantee poorer welds. These weld failures were much more apparent in deeper beams (W33 and W36) than shallower sections, such as the W18 and W21 beams used in this building. Shallower beams have been shown to have more inelastic rotational capacity before the onset of the brittle fractures than deeper sections.

Post-Northridge research has provided several options for connections that are more robust and ductile than the directly welded flange connections. So, while the older connections may possibly not perform poorly, it can be assumed with reasonable certainty that a new connection made with notch-tough weld metal and under significantly more quality control will perform better than older welded connections.

5.2.4 Effect of High Seismicity Detailing

The last item that the study addressed was the impact of present-day high seismicity detailing on the blast response of the structure. The perimeter frame was re-detailed to conform to the Special Moment Frame provisions in AISC 341 (Reference 8) without applying additional design forces. As discussed in the previous section, the columns were reoriented to face in the plane of the frame, as they would be if designed today. The analysis showed that the columns deformed more than the original building because the weak axis was loaded by the blast and the beams had greater displacements due to the loss of weak axis capacity due to the reduced beam section (from the RBS prequalified connection chosen). However, the beam-column connections and the column splices were judged to be more robust because these details are designed with specific detailing and quality control measures to remain intact as the beams yield and thus were expected to prevent a collapse. They were estimated to provide a 43% saving in floor area loss based on the worst-case estimates.

5.3 Sources of Uncertainty in Analyses

The lack of research data on the behavior of steel structures under blast loads led to uncertainty in the use of the SDOF modeling. The following issues arose:

*   The blast and impact response characteristics of the column splices found just above the second floor slab were not known. Therefore, it was assumed that the brittle behavior of those splices and lack of supplemental gravity support ability would result in zero capacity after the blast.

*   The load and resistance functions associated with the SDOF models do not accurately portray the intense, localized loading and failure of members very close in to the blast, because they were developed as simplified techniques assuming an idealized uniform loading function, identical arrival time over the member length, flexural failure modes, and uniform response to the load. The FEA and blast test provided important insight into that behavior, particularly for the members closest to the blast.

- The support conditions of the columns in the SDOF model were assumed to be fixed-fixed conditions. In FEA and test results, the deformations of the adjoining framing and the base plate condition were shown to affect the column response, which was a known limitation of the SDOF.

- The effects of large deflections in the webs and flanges in both the columns and girders, including buckling and torsion, are not well known; therefore, it was estimated that the beams' post-blast condition had zero capacity.

- The responses of the girder-to-column connections in the blast and impact of the cladding debris environment have not been tested extensively. This could have significant impact on determining column lateral support, catenary mechanisms to resist progressive collapse, and floor support.

- The blast response of the exterior girders that supported the floor slabs was not understood. The girders had limited lateral support at their top flanges through plug/puddle weld attachments to the steel decking in the floor system and the intermediate lateral supports provided by the floor purlins.

- The blast response of the floor system – a reinforced concrete slab cast over a steel deck – was not known because of lack of test data. Therefore it was estimated by the project team based on limited SDOF analyses that it had collapsed.

- Gravity load-based axial load effects on the column blast response were not tested, since they were not believed to be significant in this study.

- Ground shock response was uncertain and not considered in the evaluation.

- The ability of the floor beams and slabs to support load in damaged states through catenary resistance was not quantified through testing.

5.4 Applicability of the Results

5.4.1 Different Structural Systems

The study specifically addressed the case of upgrading a mid-1970's ordinary steel moment frame building that had been designed largely to support gravity loads so that it could resist lateral loads in an area of high seismicity (i.e., SDC D to F); subsequently, the study analyzed the building's response to a blast scenario similar to the 1995 Oklahoma City bombing of the Murrah Building. The need for significant upgrade was based on the assumption of high seismicity at a hypothetical site in San Francisco. The original frame was also subjected to a seismic re-detailing scheme that in effect upgraded its exterior frames to special moment frame capability; the re-detailed system was also analyzed for its response to an Oklahoma City bombing scenario.

The improvement in the performance of the system when an exterior BRBF upgrade was applied did confirm the FEMA 439A conclusion that the concept of applying seismic strengthening schemes to exterior frame elements can improve blast and progressive collapse resistance. This is because exterior seismic strengthening measures generally create tougher, more ductile elements along the portion of the building most susceptible to large explosive threats such as vehicle bombs. The degree to which seismic strengthening measures can improve blast resistance in structural systems other than those evaluated in this report and FEMA 439A should be determined through specific studies.

5.4.2 Different Explosion Scenarios

The 1995 Oklahoma City bombing was used as the basis for the analyses of the seismic strengthening schemes, because the estimated blast effects on the Murrah Building were fully documented, providing an experiential benchmark. That event involved the detonation of the equivalent of 4,000 pounds of TNT at a standoff distance of approximately 15 feet from a critical exterior column that supported a large transfer girder. Current design criteria, such as the IBC (Reference 38) and ASCE 7 (Reference 39), generally do not explicitly consider explosion scenarios of any intensity. Any number of other blast scenarios could also be analyzed by varying the charge weight, the standoff distance, or both. In many urban locations, it would be possible for smaller vehicles to park even closer to a building than the distance in the 1995 bombing. Each such scenario should be analyzed for its particular effects on a building.

Additionally, this study concerned itself with an exterior vehicle bomb. The effects of explosions inside the building behave differently and consequently load the structure in a different manner than exterior explosions. The applicability of the conclusions of this study to interior explosive threats would need to be evaluated separately.

5.4.3 Differences Between Seismic and Blast Demands

In the years since the Murrah Building bombing, there has been extensive debate on how well seismic structural design measures help a building to resist blast loads and progressive collapse. There are significant differences between the causes and effects of earthquakes and blasts. For response to earthquakes, less mass is better. For response to blast, however, more mass is better. Earthquake loads occur over periods of seconds, or even tenths of seconds, with subsequent cyclic response. Blast effects occur within a few thousandths of a second and generally do not result in significant cyclic response. The distribution of demands over a building is dramatically different in earthquake and blast loadings. Earthquake loads exercise the entire building, while blast loads on the order of the Murrah attack often directly involve only a relatively small portion of the building. Blast loads require energy absorption while earthquake loads require energy dissipation.

Despite these differences, earthquake-resistant design produces tough structures that can benefit both earthquake and blast resistance. The structural engineering community is already familiar with the use of earthquake-resistant structural design provisions and details contained within the model building codes. This familiarity can help engineers, regardless of the regions of seismic risk, use those provisions and details to design tougher, more robust structures, therefore reducing (while not necessarily eliminating) the structures' potential for significant damage due to an explosive hazard.

Design measures, such as those for earthquake-resistant design, do not predict structural response for all extreme loading scenarios. Rather, they produce safe and serviceable structures for the general expected loading conditions. On its surface, this study implies that modern special moment frame design and detailing might offer only marginal improvement in the blast response of this particular building. This is due to the conclusion that the original frame would have suffered limited damage. However, it must be remembered that numerous sources of uncertainty have been cited in this study. The original structure used in this study was especially robust and redundant, with the use of full moment connections on almost all the joints and also the sections chosen because the floor was not designed compositely with the slab. That robustness enhanced the survivability of the structure that was studied, and implies that the thinner, more flexurally efficient sections could be more vulnerable.

5.5   *Recommendations for Further Study*

The results of this research raise a number of important issues that could be addressed more definitively through separate studies.

As noted above, application of the results of this study to other structural systems should not be undertaken without thorough investigation. Therefore, studies of other structural systems should be undertaken.

The many sources of uncertainly cited in Section 5.3 all merit study through a program of combined experimental and analytical research if the blast and impact response of steel moment frames is to be better understood. As cited in FEMA 439A, experimental research is needed on the behavior of conventional buildings when large displacements and joint rotations occur, as would be the case in progressive collapse scenarios when members are removed or destroyed. Current SDOF analytical modeling techniques do not adequately address behavior that differs greatly from conventional small-angle flexural response and are expected to yield very conservative results. The significance of catenary action in beams, columns, and slabs is intuitively clear but not readily accounted for by calculation. Large displacement behavior, particularly in connection and joint regions, is also not well characterized. The relatively small displacements (i.e., rotations that do not engage significant catenary tension) that are typically assumed in seismic response, which have been researched more extensively, likely do not apply directly. These areas of uncertainty are present for all structural systems, but are more significant when the local response of structural steel member flanges and webs is also considered.

Impact loading effects caused by the loss of gravity load-bearing elements are not well characterized for systems that yield while resisting impact loads. The impact factor of 2.0 contained in most progressive collapse guidelines and initially used in this study applies to the increase in the force and displacement of an SDOF model resisting an impact load elastically, under small displacement. As with most progressive collapse scenarios, resistance to impact is gained through inelastic deformations and large displacement actions, both of which have been shown to have displacements greater than 2.0 times the elastic displacement of the initial load. In addition, collapse resistance is a multi-degree of freedom (MODF) response, and it is not currently known how well SDOF approximations replicate the actual response. Further study on how structures resist impact loads caused by the loss of gravity elements would be greatly beneficial to the engineering community.

To minimize the need for rigorous blast and progressive collapse analyses on conventional buildings, standard structural details that enhance blast and progressive collapse resistance should be developed. Such development requires experimental testing accompanied by analytical modeling. This detailing should be developed for major structural systems and materials that are used in construction for buildings that are three stories or more in height. Particularly in the design of simpler, regular structural systems, the use of standard details for the design of new conventional buildings could reduce the need for extensive progressive collapse analysis.

Some synergy between seismic detailing and blast and progressive collapse resistance was indicated in the results of this study, though the issue of section properties must be studied further before this can be definitively established. This apparent synergy should be investigated with the objective, if verified, of developing guidelines on tailoring seismic upgrades to maximize blast and progressive collapse resistance depending on section properties of key blast-vulnerable elements. Such guidelines could enhance the overall safety of typical buildings in seismically active regions and potentially facilitate the wider use of accepted seismic details.

In both the FEMA 439A study and in this study, the issue of floor slab response was found to be a great source of uncertainty. Just how vulnerable are floor slabs with and without underlying steel decking? Further experimental research into that vulnerability is needed, accompanied by research into more blast-resistant floor slab systems. From such research, standard blast-resistant floor slab details could be developed.

The effects of high loading rates caused by blast loading on steel connections should be researched further experimentally. Some of the past research on hardened structures can be used, but more research is needed for elements and members under high strain-rate loading.

Accompanying additional experimental research, improvements in available blast and progressive collapse response modeling procedures should be undertaken. The blast response analyses performed in the current study incorporated simplified techniques based on SDOF modeling of individual structural elements, supplemented with some finite element analysis. The SDOF models used in the study are design-conservative, based on extensive experimental research on reinforced concrete hardened structure design. The design-conservative approach involves response estimates that are generally on the high side of anticipated response. This means that the response will be predicted to be larger than expected, but will produce a safe design. Therefore, the significant improvements in predicted blast resistance with the seismic strengthening schemes in place that were reported in this study are probably less than reasonably could be expected. While the finite element analysis used in this study was consistent with the field test in its prediction of the response of the column closest to the blast, there is still much research that is needed to improve upon the FEA's ability to accurately predict blast response. This is particularly true with the behavior of structural assemblages, connections, and steel nonlinear material models at rupture. With additional experimental research, both the SDOF and FEA models can be refined to better predict structural response to blasts.

6 References

1. Federal Emergency Management Agency, *The Oklahoma City Bombing: Improving Building Performance Through Multi-Hazard Mitigation*, FEMA 277 (Washington, DC), August 1996.

2. Federal Emergency Management Agency, *Blast-Resistance Benefits of Seismic Design, Phase 1 Study: Performance Analysis of Reinforced Concrete Strengthening Systems Applied to the Murrah Federal Building Design*, FEMA 439A (Washington, DC), December 2005.

3. International Code Council, *2003 International Building Code*, CENGAGE Delmar Learning (Clifton Park, NY), February 2003.

4. Structural Engineering Institute – American Society of Civil Engineers, *Seismic Evaluation of Existing Buildings*, ASCE Standard, ASCE 31-02, (Reston, VA), 2002.

5. Federal Emergency Management Agency, *Handbook for the Seismic Evaluation of Buildings – A Prestandard*, FEMA 310 (Washington, DC), January 1998.

6. Federal Emergency Management Agency, *Prestandard and Commentary for the Seismic Rehabilitation of Buildings*, FEMA 356 (Washington, DC), November 2000.

7. Federal Emergency Management Agency, *Recommended Seismic Evaluation and Upgrade Criteria for Existing Welded Steel Moment Frame Buildings*, FEMA 351 (Washington, DC), June 2000.

8. American Institute of Steel Construction, *Seismic Provisions for Structural Steel Buildings*, ANSI/AISC 341-02 (Chicago, IL), May 2002.

9. Federal Emergency Management Agency, *Recommended Seismic Design Criteria for New Steel Moment-Frame Buildings*, FEMA 350 (Washington, DC), July 2000.

10. American Institute of Steel Construction, *Manual of Steel Construction*, Sixth Edition, New York, NY, 1968.

11. American Institute of Steel Construction, *Seismic Provisions for Structural Steel Buildings*, ANSI/AISC 341-05 (Chicago, IL), May 2005.

12. American Institute of Steel Construction, *Manual of Steel Construction, Load and Resistance Factor Design*, Third Edition (Chicago, IL), 2001.

13. Steel Deck Institute, *Diaphragm Design Manual*, Third Edition (Fox River Grove, IL), 2004.

14. Federal Emergency Management Agency, *Improvement of Nonlinear Static Seismic Analysis Procedures*, Preprint Edition, FEMA 440 (Washington, DC), February 2005.

15. Federal Emergency Management Agency, *NEHRP Recommended Provisions for Seismic Regulations for New Buildings and Other Structures, Part 1: Provisions*, FEMA 450-1/2003 Edition, (Washington, DC), 2004.

16. Federal Emergency Management Agency, *NEHRP Recommended Provisions for Seismic Regulations for New Buildings and Other Structures, Part 2: Commentary*, FEMA 450-2/2003 Edition, (Washington, DC), 2004.

17. Federal Emergency Management Agency, *Recommended Specifications and Quality Assurance Guidelines for Steel Moment-Frame Construction for Seismic Applications*, FEMA 353 (Washington, DC), June 2000.

18. Whirley, R. G. and B. E. Engelmann, "DYNA3D: A Nonlinear Explicit Three-Dimensional Finite Element Code for Solid and Structural Mechanics," User Manual, Report UCRL-MA-107254, Rev. 1, Lawrence Livermore National Laboratory (Livermore, CA), November 1993.

19. Vaughan, D. K., 1983, "Flex User's Guide," Report UG 8298, Weidlinger-Associates (Los Altos, CA), May 1983, plus updates through 2002.

20. "LS-DYNA User's Manual – Version 960," Livermore Software Technology Corporation, March 2001.

21. Attaway, S. W., et al, "PRONTO 3D User's Instructions: A Transient Dynamic Code for Nonlinear Structural Analysis," SAND98-0000, Sandia National Laboratories (Albuquerque, NM), March 1998.

22.   Magallanes, J. M., K. B. Morrill, and J. W. Koenig, "Results from Discrete Leto Tests 1 to 3 of the DTRA/GSA Steel Test Program," Karagozian & Case (Burbank, CA), IR-05-4.1, in publication.

23.   Magallanes, J. M., K. B. Morrill, and J. W. Koenig, "Discrete Leto 4: Analytic and Test Results," Karagozian & Case (Burbank, CA), IR-05-13.1, September 2005.

24.   Magallanes, J. M., K. B. Morrill, and J. W. Koenig, "Discrete Leto 5: Analytic and Test Results," Karagozian & Case (Burbank, CA), TR-05-15.1, September 2005.

25.   Magallanes, J. M., K. B. Morrill, and J. W. Koenig, "Discrete Leto 6: Analytic and Test Results," Karagozian & Case (Burbank, CA), IR-05-16.1, September 2005.

26.   Bogosian, D. D., W. Wathugala, and B. Gerber, "Fast-Running Models for Predicting Response of CMU Walls and Reinforced Concrete Columns and Beams," Karagozian & Case (Burbank, CA), TR-04-7.2, May 2004.

27.   Crawford, J. E. and L. J. Malvar, "User's and Theoretical Manual for K&C Concrete Model," Karagozian & Case (Burbank, CA), TR-97-53.1, December 1997.

28.   Malvar, L. J. and J. E. Crawford, "Recommended Static and Dynamic Properties of Steel Reinforcing Bars," TR-97-41.1, Karagozian & Case (Glendale, CA), September 1992.

29.   "Connection Test Summaries," SAC Joint Venture (Sacramento, CA), SAC-96-02.

30.   Hyde, D. W. "User's Guide for Microcomputer Programs ConWep and FunPro, Applications of TM 5-855-1," *Fundamentals of Protective Design for Conventional Weapons*, Instruction Report SL-88-1, Department of the Army, Waterways Experiment Station, Corps of Engineers (Vicksburg, MS), 1988.

31.   *Design and Analysis of Hardened Structures to Conventional Weapons Effects*, Army TM 5-855-1, Air Force AFPAM 32-1147(I), Navy NAVFAC P-1080, DSWA DAHSCWEMAN-97 (Washington, DC), August 1998.

32.   *Single-Degree-of-Freedom Plastic Analysis* (Span32), version 1.2.7.2, U.S. Army Corps of Engineers, Omaha District, Protective Design Center (Omaha, NE), 2003.

33. *Progressive Collapse Analysis and Design Guidelines for New Federal Office Buildings and Major Modernization Projects*, U.S. General Services Administration (Washington, DC), 2003.

34. Slawson, Thomas R., *Wall Response to Airblast Loads: The Wall Analysis Code (WAC)* (U.S. Army Engineer Waterways Experiment Station, Vicksburg, MS), November 1995.

35. Magallanes, J. M., R. Martinez, J. W. Koenig, "Experimental Results of the AISC Full-Scale Blast Test," Karagozian & Case (Burbank, CA), TR-06-20.2, March 2006.

36. Wong, K. "Testing of Vertical Load Carrying Capacity of a Damaged Column," University of Utah (Salt Lake City, UT), August 2006.

37. Federal Emergency Management Agency, *World Trade Center Building Performance Study: Data Collection, Preliminary Observations, and Recommendations*, FEMA 403 (Washington, DC), May 2002.

38. International Code Council, *2006 International Building Code*, CENGAGE Delmar Learning (Clifton Park, NY), March 2006.

39. Structural Engineering Institute – American Society of Civil Engineers, *ASCE 7-05 – Minimum Design Loads for Buildings and Other Structures*, ASCE/SEI 2005, (Reston, VA), 2005.

40. Unified Facilities Criteria, *Design of buildings to resist progressive collapse*, (UFC4-023-03), US Department of Defense (Washington, DC), January 2005.

41. Federal Emergency Management Agency, *NEHRP Handbook for Seismic Evaluation of Existing Buildings,* FEMA 178 (Washington, DC), June 1992.

42. Woodson, S. and Baylot, J., "Structural Collapse: Quarter-Scale Model Experiments," Technical Report SL-99-8, U.S. Army Engineer Research and Development Center (Vicksburg, Mississippi), August 1999.

43. American Concrete Institute, *Building Code Requirements for Structural Concrete*, ACI 318-02 (Farmington Hills, MI), 2002.

Appendices for FEMA P-439B

FEMA P-439B includes 7 technical appendices that are available in electronic format only (Adobe Portable Document Format, i.e., PDF) as archived on digital media attached to this report. The appendix titles are listed below: